Fire safety engineering

a reference guide

Richard Chitty
Jeremy Fraser-Mitchell

BRE FRS

OFFICE OF THE
DEPUTY PRIME MINISTER

Prices for all available
BRE publications can
be obtained from:
BRE Bookshop
151 Rosebery Avenue
London EC1R 4GB
Tel: 020 7505 6622
Fax: 020 7505 6606
email:
brebookshop@emap.com

BR 459
ISBN 1 86081 630 4

This guide has been produced by BRE under a contract placed by the Office of the Deputy Prime Minister. Any views expressed are not necessarily those of the Office.

BRE material is also published quarterly on CD

Each CD contains BRE material published in the current year, including reports, specialist reports, and the Professional Development publications: Digests, Good Building Guides, Good Repair Guides and Information Papers.

The CD collection gives you the opportunity to build a comprehensive library of BRE material at a fraction of the cost of printed copies.

As a subscriber you also benefit from a substantial discount on other BRE titles.

**For more information contact:
BRE Bookshop on 020 7505 6622**

BRE Bookshop

BRE Bookshop supplies a wide range of building and construction related information products from BRE and other highly respected organisations.

Contact:
post: **BRE Bookshop**
         151 Rosebery Avenue
         London EC1R 4GB

fax: 020 7505 6606
phone: 020 7505 6622
email: brebookshop@emap.com
website: www.brebookshop.com

Contents

Foreword

FRS, the fire division of BRE, has pioneered the development and acceptance of Fire Safety Engineering (FSE), both nationally and internationally through Standards and pre-normative research bodies such as ISO, BSI and CIB. Experts at FRS have contributed to the Standards document, BSI DD 240 *Fire safety in buildings* and its successor BS 7974 *Application of fire safety engineering principles to the design of buildings* as well as the ISO TC92 TR13387 *Fire safety engineering* report series.

As the use of engineering in support of performance-based regulation develops and becomes more complex, it is important constantly to develop and re-evaluate the knowledge base that underpins the design. Fire safety systems differ from nearly every other engineering system in a building; any faults or failures in design, implementation or maintenance may only become apparent during the emergency for which they are required.

This guide provides an introduction to the subject of fire safety engineering for both the novice and those currently involved in its practice. It attempts to explain the topics involved and the interconnected nature of the component parts in producing safe and cost-effective design.

This comprehensive guidance is the result of work at FRS funded by the Office of the Deputy Prime Minister with advice from a steering panel comprising a wide range of stakeholders from industry and the public sector.

It is expected that the guide will be of particular value to practising fire safety engineers, building control officials, fire prevention officers and indeed all those with a responsibility for fire safety.

Farshad Alamdari

Managing Director
FRS, the fire division of BRE

About the authors

Richard Chitty

Principal Consultant, Fire and Risk Engineering Centre

Richard Chitty joined BRE's fire group in 1973 conducting experimental work and developing instrumentation. Since 1982 he has been part of the fire modelling team, developing and using a range of fire models from simple calculations through to CFD techniques. He is an author of the BRE software ASKFRS and more recently has written the integrated user interface JOSEFINE for the CFD programs SOFIE, JASMINE and FDS, that also has links to the egress model GRIDFLOW and the CRISP risk assessment model.

In addition to developing and using fire models he has investigated several aspects of fire engineering, including a survey of backdraught for the Home Office in 1994 and an investigation of building separation for ODPM.

Richard frequently lectures on various Fire Safety Engineering courses.

Richard's main contributions to this reference guide are in sections G (General topics), 1 (Fire growth), 2 (Smoke spread and control), 4 (Detection/ Suppression) and 5 (Fire Service intervention). He has also reviewed the other sections.

Jeremy Fraser-Mitchell

Senior Consultant, Fire and Risk Engineering Centre

Since joining BRE in 1992, Jeremy has worked on the development and application of CRISP, a Monte Carlo model of entire fire scenarios to assess the risk to life due to fires in buildings. The basic structure of CRISP is a two-layer zone model of smoke flow for multiple rooms, coupled with a detailed model of human behaviour and movement. The CRISP program can also be used purely as an evacuation model.

In addition, he has investigated several aspects of fire engineering including a comprehensive survey of risk assessment techniques to assess fire resistance requirements, various analyses using the Home Office Fire Statistics database, and a cost–benefit analysis of residential sprinklers.

Jeremy has published extensively on fire risk assessment and evacuation modelling. He frequently lectures on various Fire Safety Engineering courses.

Jeremy's main contributions to this reference guide are in sections G (General topics), 3 (Structural fire protection), 6 (Human factors) and 7 (Risk assessment). He has also reviewed the other sections.

Acknowledgements

In addition to the two authors, this guide has been peer-reviewed by the following FRS staff: Professor Geoff Cox, Dr Suresh Kumar, Mr Brian Martin, Dr Howard Morgan, Professor David Purser, Mr Martin Shipp, Mr Nigel Smithies and Mr John Stevens. The authors would like to thank them for their input, and to thank Mr Anthony Burd and Mr Darren Hobbs of ODPM who have also reviewed the draft.

Preface

The definition of *fire safety engineering* presented by the Institution of Fire Engineers is:

'The application of scientific and engineering principles based on an understanding of the phenomena and effects of fire and of the behaviour of people to fire, to protect people, property and the environment from the destructive effects of fire.'

Other organisations, such as ISO, have published similar definitions (see, for example, ISO TR 13387).

The principal objective of fire engineering is, when an accidental fire occurs, to provide an acceptable level of safety. Often this will involve calculation or modelling of scenarios affecting all or part of the fire 'system'.

One of the characteristics of this new subject is its diversity. A fire safety engineer needs to consider chemistry (eg the behaviour of materials), physics (eg heat transfer, movement of smoke), civil, electrical and mechanical engineering, psychology (eg behaviour of people) as well as procedures used by firefighters and issues relating to management of large complex buildings. Invariably there is always something new to learn or a distant memory that needs revision.

A fire safety engineering approach will require taking a more holistic approach to a problem than may taken when using more prescriptive methods. Additional factors, that may be included implicitly in the prescritive approach, need to be considered explicitly.

It is often the interactions between different topics that cause difficulties in practical situations. Solutions to problems of a building's day to day use (ventilation, structural, security, etc.) may conflict with require-ments for fire safety. Fire safety engineering must be undertaken using a systematic approach to avoid potentially life-threatening omissions in the analysis. The flow diagram below outlines a suggested approach to the design process, following BS 7974: 2001: *Application of fire safety engineering principles to the design of buildings. Code of practice.*

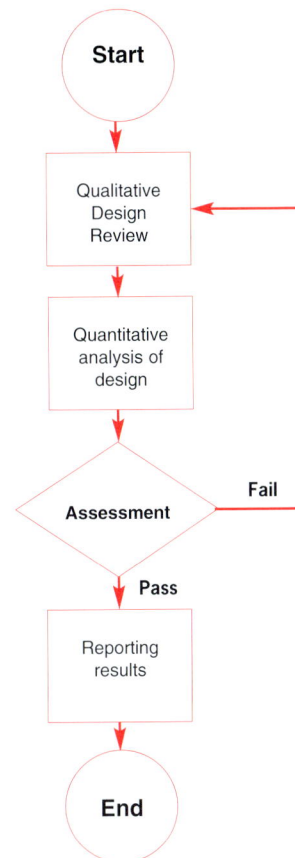

Start → Qualitative Design Review → Quantitative analysis of design → Assessment → (Fail) back to Qualitative Design Review / (Pass) → Reporting results → End

Here, particularly at the qualitative design review stage, the interaction of all the building systems are considered as well as the detailed performance of the fire protection systems. This brings people into the process who only occasionally encounter fire safety engineering or exposes a fire safety engineer to a topic that is not often encountered. There is a require-ment for these individuals to have a guide that will answer:

- 'What is this topic about?'
- 'Is there anything special to look out for?'
- 'Is it important to me as a designer or enforcer?'
- 'How does it relate to other topics?'
- 'Where can I find out more?'

Introduction

This guide is intended to be a reference for those who are new to, or occasionally encounter different aspects of, fire safety engineering. Experienced practitioners may also find it useful as a short-cut to finding information in more detailed references. The guide is not intended to be a 'one-stop shop' that will give all the information required for solving fire engineering problems during the design of a building.

The topics covered in this guide do not form an exhaustive list. Basic descriptions for a number of key topics encountered in fire safety engineering are provided and aspects that should be considered by both designer, enforcer and other responsible persons are highlighted.

If the reader requires more information then references are provided to more detailed sources. The authors have tried to limit the number of references to a few key texts that should be readily available, and have deliberately not included original articles or scientific papers that may be difficult to obtain.

The guide indicates how a particular topic is dependent on other topics and how it may influence others. To emphasise the structured approach to fire safety engineering, this guide has been organised using a number of sections similar to the sub-system approach in BS 7974. However, this grouping is not rigid, and topics in one section will have links with topics in other sections.

Detailed calculation procedures are not included; they can be found in the referenced documents.

Finally, the guide has been formatted so that it can be annotated by the owner to create a personal reference that becomes a first stop when encountering a new challenge in fire safety engineering.

Using the guide

The guide is divided into eight main sections, each covering a different aspect of fire engineering. These sections are:

G General topics
1 Fire growth
2 Smoke spread and control
3 Structural fire protection
4 Detection and Suppression
5 Fire Service intervention
6 Human factors
7 Risk assessment

Each of these sections has a one-page outline that introduces the topics covered within that section, describing how they relate to one another. There is also a one-page table that shows how topics within the section relate to the other sections within the document.

Within each section there are a number of topics, each described on a single page. The page includes a short description, any key points that designers and enforcers should look out for, references to more detailed information and how the topic links to other topics. Each topic has a unique number in the form '<section><topic>', which can be found next to the topic title, at the top-left of the page. Cross-references to each topic are shown in the form [s-t], where s is the section number and t the topic number.

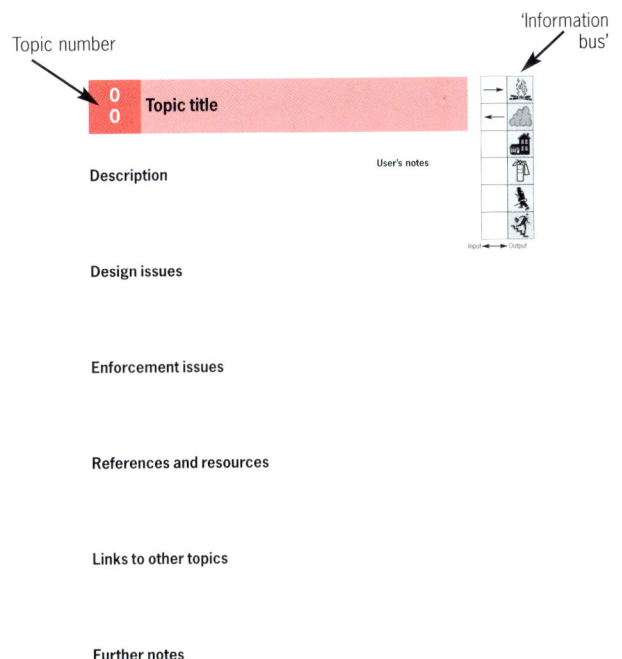

Page layout for each topic

Introduction

References and resources

The number of references has been limited to a few 'standard' books that should be easily accessible to practising fire safety engineers. The authors have endeavoured to provide at least two references for every topic. For these 'standard' references, a shortened reference is included in the text for each topic, and the full details are given in section 9, *References*. More specialised references that only apply to a single topic are given in full where they occur.

The resources given under some topics are (generally) computer software packages that enable the necessary calculations to be carried out. These are intended to be illustrative, rather than comprehensive since new software packages are being developed all the time.

Links to other topics

This highlights only the other topics with a particularly strong link. More general links to sections are shown by the 'information bus' diagrams at the top-right of each page.

Further notes

These have been provided for some topics, and may indicate 'rule of thumb' design values, or contain diagrams, etc.

Information bus

The diagram in the top right corner of the page indicates the links between each topic and the other sections of the document.

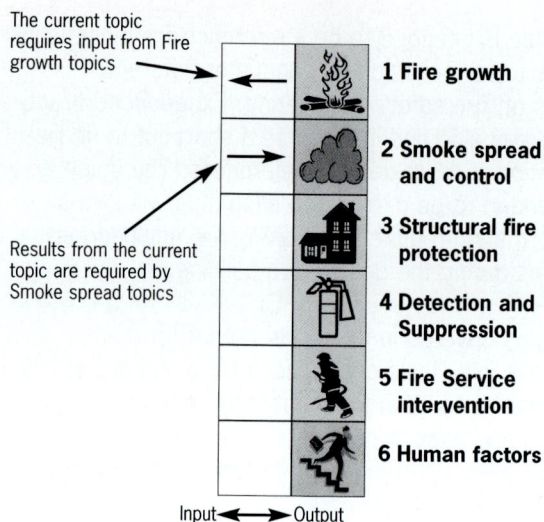

The current topic requires input from Fire growth topics

Results from the current topic are required by Smoke spread topics

1 Fire growth

2 Smoke spread and control

3 Structural fire protection

4 Detection and Suppression

5 Fire Service intervention

6 Human factors

Input ◄———► Output

The arrows to the left of the icons indicate the direction of information flow. An arrow pointing to the left means that information must come from a particular section, to provide input to the topic in question. Conversely, an arrow pointing to the right indicates that the topic provides output to be used in the appropriate section. The example above shows a topic that requires information from the fire growth section and provides data that are required by topics in the smoke spread section. The diagram can be considered as a cross-section through a data bus that flows vertically through the guide.

User's notes

Finally, space has been provided on each page for the user to add their own notes, clarifications, more details, other references, etc. In this way, the guide becomes 'customised' to each individual, and consequently more valuable.

G General topics

The topics in section G are general or basic topics that are fundamental to all the other sections. They fall into two groups:
- those that provide background to the processes of fire and heat transfer, and
- those that are at the core of the fire engineering process as described in *BS 7974: Application of fire safety engineering principles to the design of buildings*.

An understanding of some of the basic principles of combustion and heat transfer is required to use fire safety engineering effectively. This guide provides what may be considered to be a minimum requirement. The fire triangle [G-1] is an excellent aid to explain combustion, ignition and extinction. More detailed consideration of the fire growth process requires knowledge of key fuel properties [G-5]. The gas laws [G-3] are used extensively in smoke move-ment calculations and some understanding of heat transfer and the key material properties [G-4] are also required. Chapters 1–4 of *An introduction to fire dynamics* provides a good introduction to the background science required for fire safety engineering.

The process of fire safety engineering should also be understood. This guide draws on BS 7974; Part 0 of that Standard is entitled 'Guide to design framework and fire safety engineering procedures' and gives a detailed introduction. The key stages are:
- The qualitative design review (QDR) [0-6]
- The quantitative analysis
- The assessment [0-7]
- Reporting [0-8]

The QDR, assessment and reporting are included as topics in this section. Much of the rest of the guide concentrates on the quantitative analysis.

The fire triangle

Description
A sustained fire needs **three** components:
● fuel,
● oxygen, and
● heat.

This is often illustrated using the fire triangle.

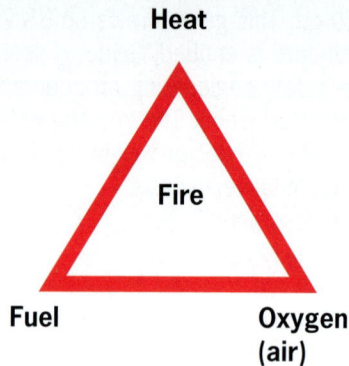

Heat

Fire

Fuel　　　　**Oxygen**
　　　　　　　　(air)

Input ◄────► Output

If two of the components are present then adding the third can result in ignition. For example, if fuel has been vaporized in a hot compartment then adding oxygen (by opening a door) can result in ignition; in this example, a backdraught or smoke explosion. The triangle can also explain extinction. Removing one of the components, such as removing heat by cooling the fuel with water, will put the fire out.

It should be noted that while air is the usual source of oxygen some fuels may contain oxygen that is released when the fuel pyrolyses (breaks down and vaporizes under the influence of heat). Strictly, the requirement is for an oxidant; some industrial processes may involve chemicals that can sustain a combustion reaction without the presence of air.

Design issues
● Basic understanding of fire behaviour

Enforcement issues
● As design

References
● *Buildings and fire*, chapter 4
● *Principles of fire behaviour*, chapter 2
● *CIBSE guide E*, chapter 9

Links to other topics
Vitiated fires [1-4], Backdraught [1-6], Effects of suppression [1-8]

Input ◄───► Output

Description

The laws of thermodynamics are some of the most fundamental concepts in physics and are relevant to all aspects of heat transfer. These are the key to many fire safety engineering problems.

The **first law** is a statement of the conservation of energy: *'You can only get out what you put in'*. In a fire most of the energy released will be translated into a rise in temperature of the gases in the compartment, compartment walls and any objects in the compartment. The increase in temperature of the gases results in their expansion.

Energy input = change in internal energy (proportional to temperature rise) + work done (proportional to change in volume)

The **second law** states that heat flow is from a hotter object to a cooler object (unless there is some external influence). Consequently, the temperature of the hotter object will fall and the temperature of the cooler object will rise until they are equal.

The **third law** states that if all the thermal motion of molecules could be removed then the temperature would be absolute zero (0 K = −273.15 °C).

A **'zeroth' law** is sometimes referred to; this states that two objects in thermal equilibrium must be at the same temperature.

Design issues

● The laws of thermodynamics are implicit in all heat transfer calculations used in fire safety engineering.

Enforcement issues

● Calculations showing excessively high temperatures (above the adiabatic flame temperature, for example, > 2000 °C) or temperatures below ambient would be the result of some fundamental error. Sudden jumps from low to high temperature remote from an obvious source of energy are also suspect.

References

● Any good physics book (eg A-level text book)

Links to other topics

Widespread links, mainly within sections 1, 2 & 3

Further notes

C P Snow provided the following definitions of the three laws:

1 You cannot win (you cannot get something for nothing as energy is conserved)

2 You cannot break even (you can't get back to exactly where you started)

3 You cannot get out of the game

Description

The relationship between the temperature, T, pressure, p, and volume, V, of gases can be written as:

$$pV = nRT$$

where n is the amount of gas (in moles) and R the universal gas constant (8.31431 $JK^{-1}mol^{-1}$). For a fixed mass of gas this can be used to relate the pressure, temperature and volume of the gas under different conditions.

Note: T is absolute temperature; the unit is the Kelvin [K] (= °C +273).

$$\frac{p_0V_0}{T_0} = \frac{p_1V_1}{T_1}$$

For a fixed mass of gas at constant pressure the equation above becomes:

$$\rho_0 T_0 = \rho_1 T_1$$

where ρ is the density of the gas. For air/smoke (at sea level atmospheric pressure) this can be reduced to:

$$\rho = \frac{351.5}{T} \ kgm^{-3}$$

This will calculate air density at any temperature. It is frequently required in smoke movement calculations.

Design issues
● These calculations are implicit in most smoke movement calculations

Enforcement issues
● As design

References
● *An introduction to fire dynamics*, chapter 1
● Any good physics book (A-level)

Links to other topics
Smoke movement topics (section 2)

Further notes
Be careful to use the correct units (Kelvin) for temperature in these calculations.

The general equation given above incorporates Avogadro's hypothesis, and the laws of Boyle, Gay-Lussac and Charles. It is also known as the 'Perfect Gas Law' or 'Ideal Gas Law'.

User's notes

Input ◄──► Output

k ρ c (kay-row-see)

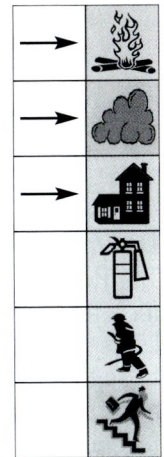

Description

Some of the most important material properties encountered in fire engineering calculations are thermal conductivity (k), density (ρ) and specific heat capacity (c).

- **Thermal conductivity (k)** is the material property that quantifies how easily heat can be conducted through a substance; it has the units of W/m/K. The U value (which is used in building energy use calculations) of an object is k/d where d is the thickness of the object.
- **Density (ρ)** is the material property that indicates the amount of substance per unit volume; it has the units of kg/m^3.
- **Specific heat capacity (c)** is the material property that quantifies the amount of energy required to raise the temperaure of a unit mass of the substance by 1 °C; it has the units of J/kg/K.

These values appear in heat transfer calculations either as the thermal inertia **kρc**, or as thermal diffusivity **k/ρc**. Good conductors of heat have high values for k and ρ but lower c, compared with insulators that have low values of k and ρ but higher c. The very high value of c for water is one of the properties that makes water an excellent extinguishing material as a small amount of water can absorb a large amount of heat with a small rise in temperature.

Each of these properties may change its value significantly with temperature.

Design issues
- The material properties k, ρ and c are used in calculations to quantify:
  - ❏ heat losses into compartment walls,
  - ❏ temperature rise of an object and ignitability.

Enforcement issues
- The properties k, ρ and c relate to the selection of materials and the ease of ignition (how quickly an object's surface temperature will rise).

References
- *An introduction to fire dynamics*, chapter 2
- *Buildings and fire*, chapter 2

Links to other topics
Heat losses to structure [2-7]

Further notes
The symbol ρ is used for both the density of solids and of gases. Care should be taken not to confuse the values in any calculation.

Input ◀——▶ Output

Assessment of designs

Description

In a fire safety engineering design following the procedures given in BS 7974, the assessment of a design requires the results of a quantified analysis to be compared with the original design criteria determined during the Qualitative Design Review (QDR).

BS 7974 gives three methods of assessing a design:

- *Comparative criteria*, where a design is shown to have a comparable level of safety to some other method [eg Approved Document B of The Building Regulations (England & Wales)].
- *Deterministic criteria*, where a specific set of safety conditions are shown to have been achieved (eg smoke layer is never below a specified height).
- *Probabilistic criteria*, where the probability of an event occurring is shown to be less than the given frequency.

Design issues

- Failure of a design to meet the assessment criteria will require the design to be modified, and the QDR and qualitative analysis cycle to be repeated until a successful design has been determined.

Enforcement issues

- The assessment method and criteria should be agreed before the quantitative design stage.

References

- *BS 7974: Application of fire safety engineering principles to the design of buildings*
- *CIBSE guide E*, chapter 2

Links to other topics

Qualitative design review (QDR) [G-6], Risk assessment (section 7)

User's notes

Input ◄───► Output

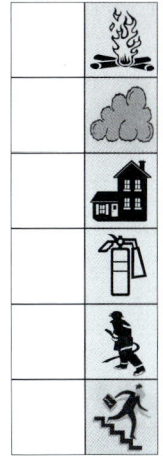

Description

The design of buildings using a fire safety engineering approach will be subject to review and approval and should be reported so that the procedures and assumptions can be readily understood by a third party. BS 7974 does not provide a specific format for reporting and presentation as a prescribed format could not anticipate all the requirements of a performance-based design. It does, however, state that the following should be included:

- Objectives of the study
- Building description
- Results of the Qualitative design review
- Quantified analysis
- Comparison with acceptance criteria
- Fire safety strategy
- Management requirements
- Conclusions
- References
- Qualifications and experience of the fire safety engineer(s)

Design issues

- Assumptions and engineering judgements should be clearly identified.

Enforcement issues

- Sufficient detail should be included so that the quantified analysis can be repeated or reviewed by a third party
- A sensitivity or uncertainty analysis [7-1] should be performed to estimate the confidence limits for the key output variables that provide the comparison against the acceptance criteria.

References

- *BS 7974: Application of fire safety engineering principles to the design of buildings*
- *CIBSE guide E*, chapter 2

Links to other topics

Qualitative design review (QDR) [G-6], Assessment of designs [G-7], Sensitivity analysis [7-1]

User's notes

Input◄──►Output

1 Fire growth

It would not be an understatement to say that the development of the fire itself is the primary determinant of the outcome of a fire scenario. 'Fire scenario' in this context means the interactions of all the subsystems.

In order to grow, the fire needs a supply of fuel, oxygen and heat (the 'fire triangle' [section G-1]). Most buildings have a plentiful supply of combustible contents, and oxygen is present in the air all around us. Fortunately, however, sources of heat sufficient to cause ignition are usually kept well away from any combustible materials so unwanted fire is a fairly rare event. Once the fire has started though, it is a different matter because the heat released by the combustion of the fuel is generally more than enough to sustain the fire, and enable it to grow larger. As the fire gets larger, it produces more heat, and so it gets larger still.

The size of the fire is commonly expressed in terms of the Heat Release Rate (HRR) (also referred to as Rate of Heat Release, RHR) [section 1-11], measured in kilowatts (kW) or megawatts (MW). The fire growth rate is the rate at which the HRR increases. This rate of increase is not usually a constant, but depends on the size of the fire. A convenient approximation for the growth phase is often given by the so-called t-squared fire [section 1-1], where the HRR is proportional to the square of the time after ignition. Other forms of 'design fire' [section 1-12] may use a steady-state fire size, or a HRR taken from experimental measurements. Small-scale measurements from a cone calorimeter [section 1-10] may be extrapolated to predict the HRR of full-scale fires, although this has pitfalls.

Most of the fire's heat is taken away by convection in the fire plume [section 1-2], the remainder (typically 30% of the total) is heat radiation from the flames. The smoke [section 1-3] produced by the fire is also carried away by the fire plume, and diluted by the air that is entrained as the plume rises. The temperature of the smoke will be sufficient to allow the formation of a buoyant layer below the ceiling.

The fire will not keep growing indefinitely — eventually it will run out of fuel or oxygen. The first way it can run out of fuel is after the item first ignited has been consumed. Whether or not the fire is able to spread beyond the first item depends on the proximity of neighbouring items, and the size the fire has reached on the first item. An important mechanism for heat transfer is radiation from the hot smoke layer beneath the ceiling. This radiation may be sufficiently intense to cause near simultaneous ignition of all fuel surfaces within the compartment — a process termed 'flashover' [section 1-5]. The HRR then increases rapidly.

A fully involved compartment fire [section 1-7] may arise either as a consequence of flashover, or the slower spread of fire from item to item until all are involved. In this case the fire size will ultimately be restricted by the availability of oxygen. The initial amount of air within a compartment will only be sufficient for the combustion of a small quantity of fuel. However, openings in the compartment walls (such as doors and windows) will allow the smoke to flow out, and fresh air to flow in at a rate that is determined primarily by the size of the openings. If the oxygen supply is sufficiently restricted, the fire is termed 'vitiated' [section 1-4]. The combustion will be less efficient, producing lower yields of the end products of combustion (carbon dioxide and water, mainly) and more smoke and intermediate products (such as carbon monoxide). In extreme cases, the temperature in the fire compartment may be sufficient for significant fuel vaporization to take place, but not enough oxygen for the combustion. If the oxygen supply is suddenly increased (eg a door opens, or a window breaks), the fuel vapours may burn rapidly when the inflow of air mixes with them, a phenomenon termed 'backdraught' [section 1-6].

The fire may be controlled or extinguished by various suppression measures [section 1-8], which either remove the heat or cut off the oxygen supply, or both. Alternatively, it will go out when it finally runs out of fuel.

User's notes

1 Fire growth

Effect **of** fire growth	Effect **on** fire growth
Smoke spread ● Fire size determines plume entrainment rate, and hence the volume and speed of smoke moving through a building ● Fire size (HRR) determines temperature of smoke layer ● Yields of combustion products, being a function of oxygen availability, affect smoke toxicity ● Soot production affects optical density	**Smoke spread** ● Radiation from hot smoke layers may enhance burning rates, or cause flashover ● If smoke completely fills a compartment, the fire will be vitiated as oxygen is consumed without replacement
Structural behaviour ● Localised heating of structural elements may occur, for example where the fire plume impinges on the ceiling ● Fully involved compartment fires transfer heat to all structural elements comprising the compartment boundary ● Structural load-bearing resistance, integrity or insulation failures may occur as a consequence of heat transfer	**Structural behaviour** ● If integrity or insulation failures occur, fire may spread into adjoining compartments
Detection/Suppression ● Detection of flames may be possible by radiation detectors ● Heat detectors/sprinklers may be activated by heat transfer from the fire plume/ceiling jet	**Detection/Suppression** ● Suppression systems will often extinguish, or at least prevent further growth of, the fire once they are activated ● The combustion chemistry may be affected, depending on the mode of suppression eg. by reduction in oxygen availability, presence of other chemicals, steam, or partial cooling
Fire Service intervention ● Rapid increases in fire growth (eg flashover or backdraught) may be hazardous to personnel	**Fire Service intervention** ● Manual suppression will extinguish or control the fire growth
Human factors ● Rapid increases in fire growth may be hazardous ● The non-linear growth of fire may give people a false impression of the time available to them ● People close to the fire may suffer from burn injuries	**Human factors** ● First-aid fire fighting may be effective in controlling or extinguishing the fire ● By opening or closing doors (or windows), people can affect the oxygen available to a fire

User's notes

Description

t-squared fire growth curves are often used with computer fire models and fire engineering calculations to approximate the growth phase of a fire. Four standard curves have been defined: Slow, Medium, Fast and Ultra fast.

Input ←→ Output

t-squared fire growth curves		
Growth curve	Time to reach ~1 MW [s]	Constant α [kW/s^2]
Slow	600	0.00293
Medium	300	0.01172
Fast	150	0.0469
Ultra fast	75	0.1876

The heat release rate at any time is $Q = \alpha t^2$ [kW] where α is selected from the above table and t is time (secs) from ignition. Typical scenarios are:

❏ Densely packed paper   Slow
❏ Traditional mattress or armchair   Medium
❏ PU mattress or PE pallets   Fast
❏ High rack storage   Ultra fast

Other values may be used for different applications and to account for the effect of fire suppression systems.

Design issues

● The t-squared curves are approximations for the growth phase of the fire. At some point the fire will stop growing and the heat release rate may become steady or fall until the fuel is consumed. The user should be aware that using a t-squared growth rate with long times will give high (possibly unrealistic) HRRs.
● The 'standard' t-squared curves are associated with typical examples (eg office fires). The use of a particular curve for an application must be justified.
● Fire area [section 1-12] should be consistent with heat output; the ratio of HRR to area typically lies in the range 250 kW.m^{-2} to 2000 kW.m^{-2}

Enforcement issues

● Fire growth curves must be representative of the appropriate fire load and fire protection measures

References and resources

● *CIBSE guide E*, chapter 9
● *The SFPE handbook of fire protection engineering*, chapter 4-13
● *Design methodologies for smoke and heat exhaust ventilation* (BR 368), chapter 3
● Some computer models require the use of t-squared heat release rate curves (DETACT-T2)

Links to other topics

Heat release rate [1-11], Fire area [1-12], Compartment fire modelling [1-9]

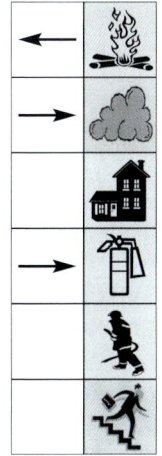

Description

A fire plume is a region of hot flowing combustion products, including flames, above a fuel source. The rising flow draws in and mixes with the surrounding air (entrainment). This reduces the velocity and temperature of the combustion products but increases the total volume flow rate. The concentration of the combustion products is also reduced due to dilution.

Fire plumes can be divided into three regions.
- Continuous flaming directly above the fuel surface, where the axial temperature is approximately constant.
- An intermediate zone where flames are present only some of the time.
- A thermal plume above the flames.

There are a number of empirical relationships (ie derived partly from experimental data and partly from theoretical principles) relating the fire HRR, fire dimensions, height of the plume and the mass flow rate of hot gases. These are for various configurations such as unrestricted plumes, plumes against walls or corners, line plumes, etc.

The plume may be deflected from the vertical, as a result of wind, air flowing into a compartment or proximity to walls. This deflection can increase the entrainment by up to 100%.

Design issues
- The correct choice of plume expression is important in the design of smoke control systems to calculate the total amount of smoke and hot gases that need to be exhausted.
- The entrainment rate depends on the height of plume rise, the heat release rate of the fire, the fire area or perimeter and any deflection from vertical.

Enforcement issues
- As design

References
- *An introduction to fire dynamics*, chapter 4
- *The SFPE handbook of fire protection engineering*, chapter 2-1
- *BS 7974: Application of fire safety engineering principles to the design of buildings*. Part 2: Spread of smoke and toxic gases within and beyond the enclosure of origin

Links to other topics
Fire area [1-12], Spill plumes [2-4], Compartment fire modelling [1-9]

Further notes
Fire plumes, deflections, etc. are handled inherently by CFD fire models, but have to be explicitly accounted for within a zone model.

User's notes

Input ◄───► Output

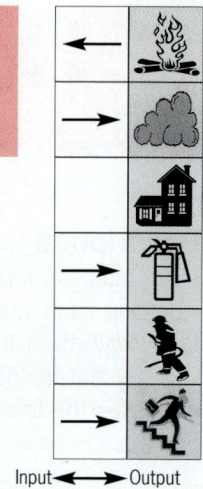

Input ◄──► Output

1
3

Smoke properties

Description

BS 7974 describes smoke as:

'The airborne products of combustion from a fire, together with large volumes of air that become entrained into them due to their motion. These combustion products can contain solid and liquid particulates, within a gaseous mass'.

The properties of the smoke depend on what is produced in the fire and on how the smoke is diluted as it moves around a building.

The most obvious property of smoke is **obscuration**; this creates a hazard to building occupants even if they are not immersed in the smoke, as they may not be able to see exits or exit signs. Smoke obscuration is measured using a number of test methods including the cone calorimeter and is usually quantified in terms of **optical density** or by an **extinction factor**.

The combustion products from the fire are a part of the smoke and may create a **toxic atmosphere** due to individual chemical species (eg carbon monoxide, or hydrogen cyanide) or a combination of species. The concentrations will depend on the product yield of the different species of the material burning and the dilution that occurs as the smoke moves through a building.

Smoke also contains heat that may present a thermal hazard to occupants and to the structure itself.

Design issues

● Numerical models can calculate the dilution of smoke as it moves away from the fire, but may have a very simple model for its production rate that may not be valid for both fuel and oxygen controlled fires. Experimental measurements of smoke production may therefore be required for the 'source term'.

Enforcement issues

● What are acceptable levels of smoke obscuration or concentration for different buildings? (see Tenability limits, section 6-7).
● The smoke layer may be above people's heads, yet still present a hazard due to heat radiation.

References

● *An introduction to fire dynamics*, chapter 11
● *The SFPE handbook of fire protection engineering*, chapters 2-13 and 3-4

Links to other topics

Tenability limits [6-7], Cone calorimeter [1-10]

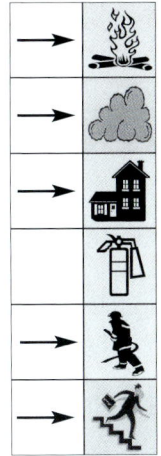

Description

Either the amount of fuel or the amount of air (oxygen) that is available will limit the burning rate of a fire. If the fire is fuel controlled then the combustion reactions can continue further to completion, achieving the maximum HRR and the combustion products will be mainly carbon dioxide and water. In practice, ideal combustion is not achieved in fires, and some fuel remains unburnt leading to soot formation and other products such as carbon monoxide.

If the fire is controlled by the air supply then many of the combustion reactions will be incomplete and the production of soot and other combustion products will be significantly increased. These are known as **vitiated fires** or **ventilation controlled fires**. After flashover, a fire will generally be ventilation controlled. The combustion products from vitiated fires, such as carbon monoxide (CO) and hydrogen cyanide (HCN), are often toxic and responsible for many deaths due to smoke inhalation.

Design issues

● The amount of air (oxygen) required for complete combustion depends on the stoichiometric ratio of the fuel.

Enforcement issues

● The specification of the correct heat release (and rate) is crucial to the fire safety design

References

● *An introduction to fire dynamics*, chapter 10
● *The SFPE handbook of fire protection engineering*, chapter 3-4

Links to other topics

Flashover [1-5], Fully developed compartment fires [1-7], Tenability limits [6-7], Backdraught [1-6], Fire triangle [G-1], Fuel properties [G-5]

Further notes

Other terms related to vitiated fires are:
● Post flashover fires,
● Fully developed fires.

Note: Pre flashover fires in tightly sealed compartments may also become vitiated as the air (oxygen) in the compartment is consumed and not replaced.

User's notes

Input ◀──▶ Output

1 5 Flashover

Input ←——→ Output

Description

Flashover is a rapid change from a localised fire in a compartment to a fire involving all the surfaces of combustible objects in the compartment.

As a fire in a compartment grows, the temperatures of the compartment walls and the hot gas layer increase, radiating heat back to the fire and surrounding combustible materials. This increases the HRR of the fire and causes it to spread. In addition, other objects in the compartment are heated. This is a classic positive feedback process; the larger the fire, the hotter the compartment temperature and the more the fire size grows. At some point, the non-burning items in the compartment reach their own ignition temperature and start to burn. The growth only stops when the fire is limited by the amount of oxygen that can enter the compartment or all the combustible materials are involved. After the flashover transition has occurred, vaporized fuel that cannot burn inside the compartment will burn outside the openings (doors or windows).

The flashover transition occurs when the upper smoke layer temperature has reached about 550 °C, or the radiation to the floor is above about 20 kW/m^2.

User's notes

Design issues

● Flashover is a natural part of fire growth in a compartment where there is sufficient fuel and ventilation. This cannot be prevented by specific design features, other than limiting the amount of fuel or by using a suppression system.
● Action can be taken to delay the effect by:
  ❑ increasing the distance between potential fuel items.
  ❑ limiting the supply of air to the compartment; however, this may create a vitiated fire and conditions that could lead to a backdraught.

Enforcement issues

● As design

References

● *An introduction to fire dynamics*, chapter 9
● *Buildings and fire*, chapter 5
● *CIBSE guide E*, chapter 9

Links to other topics

Backdraught [1-6], Fully developed compartment fires [1-7]

Further notes

If a compartment reaches flashover then life will not be tenable there and structural damage can be expected.

Backdraught

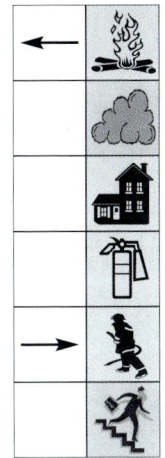

Input ←→ Output

Description

If a fire occurs in a compartment that has restricted ventilation, after a short time there will be insufficient oxygen to sustain combustion. This may extinguish the fire. However, the compartment temperature may be high enough for the fuel to continue vaporizing or smouldering. This may create an atmosphere with a high concentration of flammable gases. Changing the ventilation under these conditions (by opening a door or by a window breaking) will allow air to enter the compartment and create a mixture of gases that will ignite if there is a suitable ignition source, such as a glowing ember from the original fire. This will result in an 'explosive' event, probably with flaming through the new ventilation opening.

User's notes

Design issues
● See enforcement issues

Enforcement issues
● Backdraught concerns those (usually firefighters) entering and searching a well sealed building (eg basements, or unfenestrated buildings) if a fire is suspected.
● Ventilation techniques used by firefighters, such as Positive Pressure Ventilation (PPV), can reduce the risk of a backdraught. Suppression techniques such as Offensive Fog Attack (see www.firetactics.com) may also be effective.

References
● *An introduction to fire dynamics*, chapter 9
● *CIBSE guide E*, chapter 9

Links to other topics
Flashover [1-5], Fire triangle [G-1]

Fully developed compartment fires

Input ◄────► Output

Description

After flashover, all the compartment's combustible surfaces pyrolyse (heat causes solid substances to break down creating flammable gases). The gaseous fuel created by pyrolysation burns when it encounters air to produce high compartment temperatures, typically 1000–1100 °C, but possibly as high as 1300–1400 °C (sufficient to fuse the surface of bricks). During this phase of the fire, building elements (walls, beams, columns, etc.) may become hot enough for them to 'fail' either by losing their structural strength or allowing fire spread into adjacent compartments.

The production of gaseous fuel is usually so high that the air entering the compartment is insufficient for complete combustion. This results in an increased production of smoke and toxic products such as carbon monoxide. The smoke will contain unburnt fuel gases, which will burn as the smoke flows out of the compartment, encountering fresh air.

User's notes

Design issues

- Several calculation methods are available for the estimation of HRR and temperature in a fully developed compartment fire.
- These calculations can lead to estimates of flame lengths outside compartment windows, etc.

Enforcement issues

- Fully developed fires are of most significance for the impact of the fire on the structure or for tenability conditions elsewhere in a building; conditions for people in the fire room will be completely untenable.

References

- *An introduction to fire dynamics*, chapter 10
- *The SFPE handbook of fire protection engineering*, chapter 3-6
- *Buildings and fire*, chapter 5
- *Principles of fire behaviour*, chapter 9
- *CIBSE guide E*, chapter 9

Links to other topics

Flashover [1-5], Fire severity [3-2], Fire resistance [3-1]

Further notes

This phase of fire development is also referred to as a:
- post flashover fire,
- fully involved fire,
- ventilation-controlled fire.

If the compartment openings are very large, oxygen availability may not be the limiting factor for fire size. Instead, this is determined by fuel load density ($kg.m^{-2}$)

Effects of suppression

Input ◄──► Output

Description
The fire triangle shows that fires can be suppressed by removing the fuel or air (oxygen) or heat.

Water sprays from sprinklers provide an effective mechanism for removing heat. However, it can be difficult to ensure that water is delivered to all parts of the fire (eg fire on high rack storage systems). The spray may not completely extinguish the fire but should control it by cooling adjacent fuel and reducing fire spread. Combustion inhibitors such as halons terminate the combustion reaction (by chemical means) and hence reduce the production of heat.

Oxygen can be removed from a fire either by stopping the air-flow into the fire (closing a compartment or applying a fire blanket) or by displacing the oxygen in the atmosphere by an inert substance such as carbon dioxide, nitrogen or steam.

Removing the fuel may involve turning off a flow of gas or flammable liquid. In the case of a gas/oil well blow-out this may require drilling a new shaft to bypass the burning well-head. It may also be possible physically to remove unburnt fuel from near the fire (eg digging out a burning landfill site or moving cars in a car park).

User's notes

Design issues
● A water spray will also cool the hot gas layer and may make a natural smoke exhaust system less efficient. However, good design practice can compensate for this.

Enforcement issues
● Some suppression systems may have adverse side-effects. For example, (accidental) discharge of inert gas systems in a closed room can suffocate occupants.
● Halons are environmentally unfriendly, and production of halons is now banned by the Montreal Protocol. Halon systems can only be specified if they use recycled gases.
● Cooling of the hot smoke by a water spray may lead to a loss of buoyancy and mixing of the smoky layer with clear air beneath, thus causing poor visibility in escape routes.

References
● *CIBSF guide E*, chapters 8, 9
● *Design methodologies for smoke and heat exhaust ventilation* (BR 368), chapter 5

Links to other topics
Fire triangle [G-1], Backdraught [1-6], Sprinklers [4-6], Other suppression systems [4-7]

Compartment fire modelling

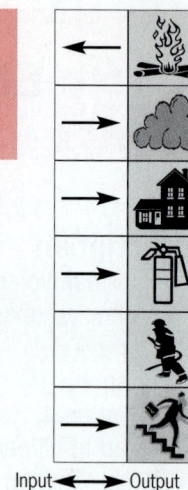

Description

A number of mathematical models exist to predict the conditions in a compartment (and adjoining rooms) when a fire occurs.

Zone models usually assume that a fire fills a homogeneous hot gas layer under the ceiling of the compartment. As the layer deepens, it spills under the lintel of open doors and windows (inverted bath tub analogy). Empirical equations calculate the amount of smoke entering the hot layer from the fire plume and leaving through the openings. Zone models include many simplifications and must be used with caution outside the range of their original application.

The more advanced **computational fluid dynamics (CFD)** models include fewer simplifications, though still making some approximations to deal with turbulence, radiation, etc. These divide the volume of interest into a large number of small cells, and at each cell solve fundamental equations for the movement of fluids. This gives very detailed results for the movement of smoke and hot gases in a building. CFD models are more complex to use than zone models and may require long computing times (several days) for fire problems.

User's notes

Design issues
- Be aware of the assumptions in any compartment model and justify them for the particular application.
- Several runs of a model may be required, to test sensitivity to input parameters.
- The ratio of HRR to fire area typically lies in the range 250–2000 $kW.m^{-2}$.
- In CFD models, representing the fire as a volumetric heat source may give unsafe predictions.

Enforcement issues
- Strange or unexpected results from models require an adequate explanation from the user. For example, the prediction of the 'trench effect' in the King's Cross fire investigation was supported by small and large scale experiments.

References and resources
- *The SFPE handbook of fire protection engineering*, chapters 3-5, 3-7, 3-8
- *CIBSE guide E*, chapter 9
- BRE Digest 367 — *Computer fire modelling*. 1991
- Examples of computer models are CFAST (zone), JASMINE (cfd).

Links to other topics
Flashover [1-5], SHEVS [2-11], Available Safe Egress Time [2-2]

Further notes
Remember! With any computer model, no matter how impressive its graphical outputs: *Garbage in = Garbage out.*

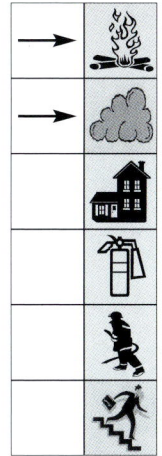

Cone calorimeter

Description

The cone calorimeter is a small-scale instrument that measures heat release rate of materials under a wide range of conditions, using the oxygen consumption technique. A square sample 100×100 mm is exposed to the radiant flux of an electric heater. The heater has the shape of a hollow truncated cone (hence the name of the instrument) and is capable of providing heat fluxes to the specimen in the range $10–110$ kW·m^{-2}. Tests can be conducted in the horizontal and vertical orientation. It is possible to enclose the instrument so that materials can be conducted at different oxygen concentrations.

In addition to HRR, the cone can provide:
● effective heat of combustion,
● smoke density,
● time to sample ignition.

Other properties such as heat of vaporization (H_{vap}), conductivity (k) and specific heat capacity (c) can be calculated.

There have been studies comparing the data collected by the cone with data from large-scale experiments.

Design issues
● Selection of materials

Enforcement issues
● As design

References
● *An introduction to fire dynamics*, chapter 9
● *The SFPE handbook of fire protection engineering*, chapter 3-3
● http://www.wpi.edu/Academics/Depts/Fire/Lab/Cone/

Links to other topics
Compartment fire modelling [1-9], kρc [G-4], Fuel properties [G-5]

Further notes
Much larger furniture calorimeters and room calorimeters use the same principle of oxygen consumption to estimate the HRR. For many materials, the heat of combustion is about 13 100 kJ per kg of oxygen consumed.

User's notes

Heat release rate (HRR)

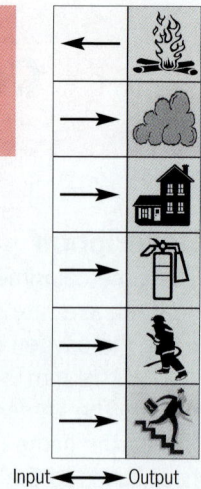

Input ←——→ Output

Description

Simply defined, Heat release rate (HRR) quantifies 'how big the fire is' and has been described (by Babrauskas, *Fire Safety Journal* 1992: **18**: 255–273) as the most important variable in fire hazard. It may also be referred to as the rate of heat release (RHR).

HRR is the rate that potential chemical energy in the fuel is converted to thermal energy by the combustion process. This is measured in Watts. In the case of fires, due to the high energy release, this will usually be kiloWatts (kW) (thousands of Watts) or MegaWatts, MW (millions of Watts). HRR is given the symbol, Q, and is related to the burning rate of the fuel, m_b, by the heat of combustion, H_c, (a property of the fuel) and the combustion efficiency, χ (how efficiently fuel mixes with oxygen: the typical value of χ for cellulosic fuels is 0.8) [section G-5].

$$Q = \chi\ H_c\ m_b$$

This provides the total thermal output of the fire. Some calculations, especially those for smoke control require that the convective and radiative components are considered separately. For a very smoky fire, up to 50% of the total heat released may be lost as radiation from the fire plume.

Design issues

● Most design calculations will be sensitive to the value of HRR, and the fraction emitted by radiation.
● The ratio of HRR to fire area typically lies in the range 250–2000 kW.m^{-2}.

Enforcement issues

● The HRR and area are key components of the characterisation of the design fire for a particular scenario, and most subsequent calculations will be dependent on these values.

References and resources

● *Introduction to fire dynamics*, chapter 4
● *The SFPE handbook of fire protection engineering*, chapter 3-1
● BRE Design fire database CD

Links to other topics

Fire area [1-12], Design fire [1-13], t-squared fire growth curves [1-1], Fuel properties [G-5]

Further notes

Note: ensure that total and convective HRR are used correctly in different calculations.

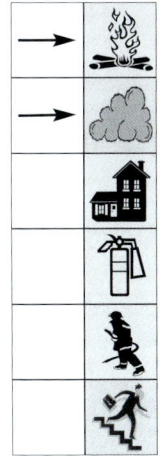

Fire area, Q/A, Q*

Description

In addition to specifying the correct HRR, it is important to specify the correct fire area. A large heat release over a small area is representative of a jet fire such as that from a broken gas pipe. Conversely, a small heat release over a large area will be more like a grassland fire. This can be checked by calculating the HRR per unit area (Q/A) which should normally be in the range 250–2000 kWm^{-2} for most materials found in buildings.

Alternatively, the non-dimensional HRR Q* (Q-star) can be calculated. This is the ratio of the momentum of the fire gases to their buoyancy and is calculated from:

$$Q^* = \frac{Q}{\rho_0 \, T_0 \, c_p \, g^{1/2} \, D^{5/2}} = \frac{Q}{1110 \, D^{5/2}}$$

where Q is the total HRR of the fire (kW) and D is a characteristic dimension of the fire (m). For fires in buildings, Q* will usually be greater than 0.1 and less than 2.5. Outside this range, most of the generally used equations for plume entrainment and flame height will be invalid.

Design issues

● Q/A and Q* should be in the ranges indicated above unless special fire conditions (such as a leak from a high pressure gas pipe) are being examined when alternative plume calculations may be required.
● Using incorrect values of Q/A or Q* will give incorrect temperatures and rates of entrainment in the fire plume.

Enforcement issues

● As design

References

● *Introduction to fire dynamics* (chapter 4)
● *BS 7974: Application of fire safety engineering principles to the design of buildings*. Part 2: Spread of smoke and toxic gases within and beyond the enclosure of origin
● *SFPE handbook of Fire Protection Engineering* (chapter 3-8)

Links to other topics

Design fire [1-13], t-squared fire growth curves [1-1], Heat release rate [1-11], Fire plume [1-2]

Further notes

Q* is an advanced topic for what is intended to be an introductory guide. However, it is important to appreciate the difference between buoyant fires and jet fires so that the correct calculations can be applied if required.

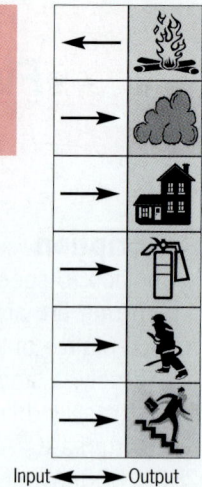

Input ◄──► Output

Description

The specification of a design fire will relate to the particular scenario being considered and will depend on:
● the type of combustible materials, and their distribution,
● potential ignition sources, and
● ventilation conditions.

A design fire will be characterised in terms of:
● HRR,
● fire area,
● production rates of products and smoke.

While HRR data exist for many materials, in practice a fire scenario will include a mixture of materials and it may be difficult to assign heat release data with confidence. Statistical analysis of fires in similar occupancies and the use of experimental fires under calorimeters may be required to support the selection of a particular design fire.

Design fires may have a constant HRR (steady state) or a time-varying HRR (transient).

Design issues
● Most fire safety engineering calculations follow from the specification of the design fire and are strongly dependent on it.
● The determination of design fires is a critical part of the role of the Qualitative Design Review team in a fire safety engineering design.

Enforcement issues
● Steady-state fires may be assumed to encompass all fires up to the size of the design fire for an indefinite time.
● Transient fires with a realistic growth phase are required for a reasonable calculation of time to detection and values of Available Safe Egress Time (ASET).

References and resources
● *BS 7974: Application of fire safety engineering principles to the design of buildings.* Part 1: Initiation and development of fire within the enclosure of origin
● *CIBSE Guide E*, chapter 9
● BRE Design fire database CD

Links to other topics
Heat release rate [1-11], Fire area [1-12], t-squared fire growth curves [1-1], Worst case scenario [7-8]

User's notes

1/14 Cables

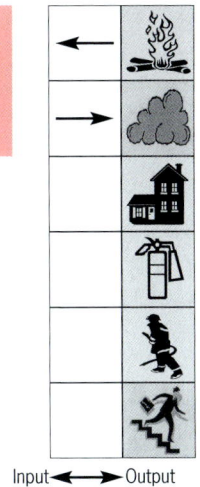

Input ◄──► Output

With the development of new technology, the numbers of communication cables in buildings have increased rapidly over the last decade. Cables can now create a significant fire load in voids within a building. Over a period of time, the number of cables can increase further as the systems are upgraded and old cables are left in situ.

Building managers should be aware of the presence of communication cables in hidden voids, especially where potential ignition sources, such as transformers for low voltage lighting circuits, may be present. *Note:* A small fire that destroys part of a computer network could result in a significant loss of business to the company.

Power circuits that are required to operate in the event of a fire should be suitably protected (*BS 6387: Specification requirements for performance requirements for cables required to maintain circuit integrity under fire conditions*), and routed through parts of the building where the fire risk is lowest.

User's notes

Design issues
● Where cables pass between compartments and through cavity barriers, adequate fire stopping must be used.
● If cables make a significant contribution to the fire load, this should be accounted for in the design.
● Communication cables with a high fire performance are available for use in hidden voids as part of a fire safety engineering design.

Enforcement issues
● The contribution of cables to a fire in a fire safety engineered building may be significant, and therefore the materials they are made of may need to be considered.

References
● The Building Regulations (England & Wales) Approved Document B (section 6.38)
● *The LPC design guide for fire protection of buildings*

Links to other topics
Heat release rate [1-11], Design fire [1-13]

2 Smoke spread and control

In building fires, smoke will usually move from the location immediately surrounding the fire to other parts of the building creating a threat to the occupants. The time taken between the ignition of the fire* and the onset of life-threatening conditions is the maximum time that occupants have to move to a place of safety. This is often referred to as the Available Safe Egress Time or ASET [section 2-2].

There is a range of methods, of varying complexity, available to calculate smoke movement. These methods include calculating the effects of:
● buoyancy of smoke/fire gases [section 2-1],
● turbulent mixing and entrainment (dilution) [sections 2-3, 2-4],
● stack effect [section 2-5],
● ventilation systems in the building, and consequent air movement [sections 2-11, 2-12],
● wind [section 2-6],
● thermal radiation [section 2-7],

and provide information about:
● mass flow rate of smoke,
● temperature of smoke,
● velocity of smoke and hot gases,
● volume of smoke,
● optical density of smoke,
● concentration of toxic gases.

From these, the onset of hazardous conditions at different locations in the building can be identified.

Features can be introduced into a building to remove and to control the spread of smoke. These may be in the form of barriers and screens [section 2-9] to contain the smoke in reservoirs [section 2-8], exhaust systems [sections 2-10, 2-11] to remove the smoke, and pressurisation systems [sections 2-13, 2-14] to prevent smoke entering sections of the building. These methods are collectively known as smoke control systems.

A smoke control system will have a major impact on ASET for the overall design of the building and must be considered at an early stage of design. Novel building concepts may require novel smoke control systems and it is essential that these can be shown to provide an adequate ASET. This may often be demonstrated using calculation methods (computer simulations [section 1-9]) or hot smoke tests [section 2-15].

---

*There has been considerable philosophical discussion about whether the ASET should be measured from ignition, or from when the fire is first detected. However, if ASET is measured from ignition (as defined in, for example, BS 7974), then the Required Safe Egress Time (RSET [section 6-1]) must also be measured from ignition, ie it should include a component for the time required to detect the fire.

User's notes

2 Smoke spread and control

Effect **of** smoke spread	Effect **on** smoke spread	
Fire growth ● Radiation from hot smoke layers may enhance burning rates, or cause flashover ● If smoke completely fills a compartment, the fire will be vitiated as oxygen is consumed without replacement	**Fire growth** ● Fire size determines plume entrainment rate, and hence the rate of smoke layer volume increase ● Fire size (HRR) determines temperature of smoke layer ● Yields of combustion products, being a function of oxygen availability, affect smoke toxicity ● Soot production affects optical density	
Structural behaviour ● Smoke layers will transfer heat to the structure ● Preheating of adjoining compartments may make their failure more rapid when fire spreads to them	**Structural behaviour** ● Compartmentation can restrict the movement of smoke around the building ● If integrity failures occur, smoke will spread into adjoining compartments	
Detection/Suppression ● Heat detectors/sprinklers may be activated by heat transfer from the smoke layers. ● Optical/ionisation detectors may also be triggered when the smoke layers reach them	**Detection/Suppression** ● Suppression systems will usually extinguish, or at least prevent further growth of, the fire once they are activated, and hence reduce smoke production ● Sprinkler sprays may cause downdrag of smoke layers ● Sprinkler sprays may cause cooling of smoke layers	
Fire Service intervention ● Reduction of visibility caused by smoke will hamper fire service operations within the building	**Fire Service intervention** ● Manual suppression will extinguish or control the fire growth, and hence reduce smoke production ● Fine sprays of water into hot smoke layers can cool them down ● Breaking windows or creating openings in the roof can remove smoke (fire venting)	
Human factors ● Toxic products of combustion contained in smoke layers are the primary cause of death/injury in fires ● People may be incapacitated by heat, either by immersion in smoke or by radiation from hot layers above them ● Obscuration by smoke hinders or prevents egress ● Irritant smoke enhances the effects of obscuration	**Human factors** ● By opening or closing doors (or windows), people can affect the movement of smoke around a building ● If doors to pressurised areas are opened for any length of time, the pressurisation will be lost	

Buoyancy

Description

It is commonly known that 'hot air rises'. This is due to the buoyancy force caused by the difference in density between the hot air and its surroundings. It is this effect that drives smoke to rise in a fire plume and to form a layer under a ceiling.

As the temperature difference between the smoke and its surroundings falls, the buoyancy force reduces and other forces acting on the gases may begin to dominate. For example, as smoke rises it cools due to dilution, and its temperature (and density) will approach that of the surrounding air. Viscous (friction) forces will then begin to dominate and the smoke will not rise any further, but may start to stratify to form a layer before reaching the ceiling. This may occur in an atrium space, due to the air temperature increasing with height under normal conditions.

Design issues
● A buoyant hot gas layer is a fundamental assumption in the design of most smoke control systems, eg the Smoke and Heat Exhaust Ventilation System (SHEVS) approach.

Enforcement issues
● If the gas temperatures are low then other air movement (eg wind) may disturb a smoke layer and the (ideally) clear layer below may become contaminated.

References
● *An introduction to fire dynamics*, chapter 11
● *Design methodologies for smoke and heat exhaust ventilation* (BR 368), chapter 2

Links to other topics
Gas laws [G-3], Smoke spread (section 2), Sprinklers [4-6]

User's notes

Input ◄──► Output

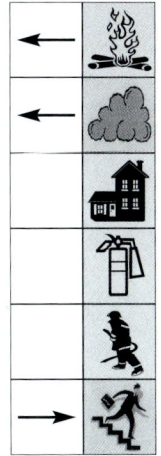

Available Safe Egress Time (ASET)

Description
The Available Safe Egress Time (ASET) is the time between the start of the fire and the onset of conditions that create a hazard to the occupants.

The concept of ASET was originally intended for an effectively closed compartment containing a fire. The concept has been broadened to include rooms not containing a fire; in these cases, calculations must account for the rate of smoke flow between rooms as well as the filling time for each room.

ASET is used in comparison with the Required Safe Egress Time (RSET), ie the time for occupants to escape. The inequality should be ASET > RSET.

Design issues
● ASET depends on the fire scenario chosen

Enforcement issues
● ASET and RSET are not single deterministic values, but will have probability distributions associated with them. Analysis must consider the safety factor, not just that ASET should be greater than RSET.
● Either ASET or RSET must account for the time delay between ignition and the fire first being detected.

References and resources
● *The SFPE handbook of fire protection engineering*, chapter 3-10
● *Buildings and fire*, chapter 12
● *BS 7974: Application of fire safety engineering principles to the design of buildings*
● Various fire simulation models calculate ASET using either zone or CFD approaches (eg CFAST, JASMINE, etc.).

Links to other topics
Required Safe Egress Time [6-1], Compartment fire modelling [1-9], Safety factors [7-5]

Further notes
An early computer model from the National Institutes of Standards and Technology (USA) was called ASET. A simplified version called ASET-B was one of the first models available for the desktop PC.

The definition of ASET is in terms of time since ignition, but it may be more helpful to calculate the time difference between the first warning being given, and the onset of untenable conditions. Ensure that ASET and RSET are expressed on a common time basis.

Input ◄──► Output

Description

If a plume from a fire rises to reach a ceiling then a relatively rapidly moving shallow layer (the ceiling jet) forms below the ceiling at the point of impingement.

In the ceiling jet region, the gas temperature and velocity is related to the heat release rate (HRR), plume height and distance from the impingement point. These relationships normally only apply to an unbounded, radially symmetric region, and do not hold once the ceiling jet has reached the walls of the compartment or corridor.

Design issues

● The response of heat and smoke detectors under a ceiling should take account of the formation of a ceiling jet where the local temperatures will be different to the properties of the hot gas layer.

Enforcement issues

● Characterisation of the ceiling jet is essential for calculating the operation times of detectors and sprinklers.

References and resources

● *An introduction to fire dynamics*, chapter 4
● *The SFPE handbook of fire protection engineering*, chapter 2-2
● *BS 7974: Application of fire safety engineering principles to the design of buildings*. Part 2: Spread of smoke and toxic gases within and beyond the enclosure of origin
● Some zone models such as DETACT assume the form of the ceiling jet whilst CFD models predict it

Links to other topics

Detector location [4-5], Sprinklers [4-6]

Further notes

The diagram shows a slice through the typical geometry of a ceiling jet, unbounded by corridor or compartment walls. The plume height of rise is 'H' (m); broad arrows show the direction of smoke movement.

Input ◄——► Output

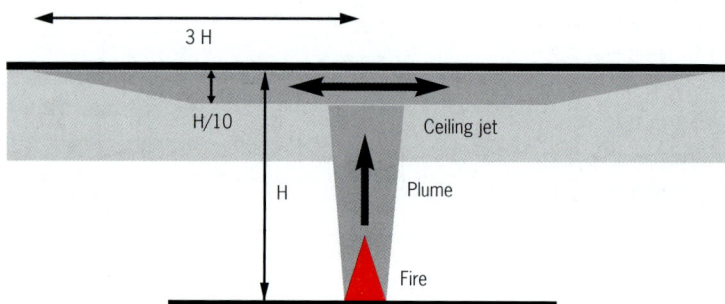

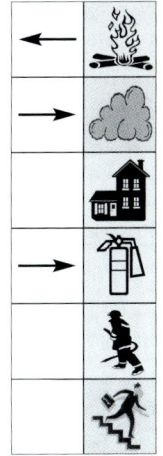

Spill plumes

Description

The rising plume of hot gases over the fire is not the only place where smoke can be diluted. Long 'line' plumes can develop where smoke spills from under a balcony (like an inverted waterfall) or past the top edge of an open window.

Design issues

- Spill plumes occur in atria, shopping malls and other geometrically complex buildings.
- Air may be entrained on one or two sides depending on whether the rising plume is attaching to a vertical surface.
- A spill plume significantly affects dilution of smoke and thereby sizing of extract systems.
- The plume width may be controlled by the use of channelling screens.

Enforcement issues

- The identification of and calculation for spill plumes is essential for the estimation of ASET and sizing of extracts in a Smoke and Heat Exhaust Ventilation System (SHEVS). Where the geometry is complex, calculations may need to be done by CFD simulations, and perhaps verified using hot smoke tests.

References and resources

- *Design methodologies for smoke and heat exhaust ventilation* (BR 368)
- *BS 7974: Application of fire safety engineering principles to the design of buildings*. Part 2: Spread of smoke and toxic gases within and beyond the enclosure of origin.
- *CIBSE guide E*, chapter 9
- CFD computer models

Links to other topics

Fire plume [1-2], Available Safe Egress Time [2-2], SHEVS [2-11]

User's notes

Input◀──▶Output

Description

Just as pressure under water varies with depth, so pressure in the atmosphere varies with height. The variation of pressure inside a building will not necessarily follow the changes in external pressure due to the temperature difference between the inside and the outside of the building. There will be a height where the internal and external pressures are equal; this is referred to as the **neutral pressure plane**. If the temperature inside the building is higher than outside, then below the neutral pressure plane the pressure inside is less than outside and air will flow into the building, while above the neutral pressure plane air will flow out of the building. The flow directions are reversed if the internal temperature is lower than the external temperature.

A fire will cause its own local pressure changes in the building. Combining pressures due to the fire and stack effect can be used to determine the direction of smoke movement at different heights in the building.

Design issues

● The presence of the stack effect and location of the neutral pressure plane is an important consideration in the design of smoke control systems in tall buildings.

Enforcement issues

● As design

References and resources

● *The SFPE handbook of fire protection engineering*, chapter 4-12
● *Buildings and fire*, chapter 9
● CFD computer models

Links to other topics

Buoyancy [2-1], Effects of wind [2-6], SHEVS [2-11]

Further notes

The stack effect is included implicitly in CFD models.

The pressure difference due to the stack effect is given by:

$$P_{inside} - P_{outside} = g.h.(\rho_{outside} - \rho_{inside})$$

where ρ is the air density (kg.m^{-3}), g is acceleration due to gravity (9.81 m.s^{-1}) and h is the height (m) above the neutral plane (below the neutral plane, h is negative).

User's notes

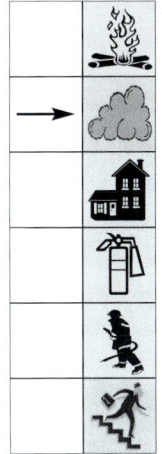

Description

The flow of air around a building due to wind can create unexpected and sometimes undesirable effects on smoke ventilation systems. This is because for some wind directions and velocities the flow through extracts or replacement air inlets may be enhanced, while under different conditions it may be impaired. This is quantified using a pressure coefficient C_p that is calculated or measured over the surfaces of the building. A negative value indicates that the wind pressure would induce an outflow from the building at an opening, while a positive value indicates an inflow.

Clearly, it is advantageous to locate extracts in regions where C_p is negative and inlets where it is positive. However, this is difficult to achieve for all wind conditions. Wind sensing devices may be installed to select the optimum use of vents for particular conditions, but this is not common practice.

User's notes

Design issues

● Wind will significantly affect the performance of natural (buoyancy-driven) smoke ventilation systems.
● The wind pressure coefficient may be found by calculation, measurement on the completed building or using a wind tunnel model
● Air flows also depend on the conditions inside the building (which will change as a consequence of these external air flows).
● The largest (or most negative) values of C_p determine the dominant openings, ie those that have the most influence on the air flows. If C_p is close to zero, flows may change direction as the system as a whole reaches equilibrium.

Enforcement issues

● As design issues

References and resources

● *Design methodologies for smoke and heat exhaust ventilation* (BR 368), chapter 10
● *Buildings and fire*, chapter 9
● BRE Digest 390 — *Wind around tall buildings*. 1994
● CFD models, wind tunnels

Links to other topics

Stack effect [2-5], SHEVS [2-11], Buoyancy [2-1]

Further notes

The pressure difference is proportional to the square of the wind speed. Small changes in wind speed can have a large effect.

Input ◄──► Output

Description
When hot gases from a fire come into contact with parts of the building structure, heat is transferred from the gas to the structure. This cools the fire gases and warms the structure.

Cooling the fire gases will increase the density of the fire gases and consequently reduce the buoyancy force that causes the fire gases to rise or stay in a stratified layer. In an extreme case, sufficient heat may be lost by the smoke to the structure so that the smoke destratifies and mixes into the air at low level creating a hazard to occupants.

Heating of the structure by the fire gases usually only presents a problem in fully developed fires of a prolonged duration, when the fire resistance properties of the material become important (eg loss in structural strength of steel at approximately 600 °C). Local 'hot spots' in the structure may also occur as a result of direct flame impingement, or radiation. However, attention may need to be paid to failure temperatures of fixings (eg for smoke curtains) and to glazing which may fail due to differential expansion between exposed and protected areas.

User's notes

Design issues
● Heat losses to the structure will reduce the temperature of hot gas layers, and hence their volume and buoyancy.

Enforcement issues
● As design

References and resources
● *The SFPE handbook of fire protection engineering*, chapter 2-14
● *Structural design for fire safety*, chapter 3
● CFD models that include heat transfer to solid objects

Links to other topics
Compartment fire modelling [1-9]

Further notes
The rate of heat transfer, Q, is by:
● **convection** $Q = h A (T_g - T_w)$
● **radiation** $Q = \sigma \varepsilon A (T_g^4 - T_w^4)$

where A = area of the wall, T_g = gas temperature, T_w = wall temperature, $\sigma = 5.67 \times 10^{-8}$ (Stefan's constant), ε = net emissivity, h = convective heat transfer coefficient (h = 5 ~ 25 $Wm^{-2}K^{-1}$).

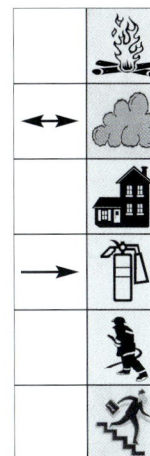

User's notes

The smoke reservoir is basically a volume at high level in a building designed to trap and hold smoke. It is intended to prevent the spread of smoke to other parts of the building beyond the reservoir. The smoke may be extracted from the reservoir at a rate matching the supply rate to provide an indefinitely long Available Safe Egress Time (ASET), or the time taken to fill the reservoir may provide an adequate ASET. The reservoir may be formed by the building structure at ceiling level; this can be enhanced by fixed or automatic smoke curtains. Fixed smoke curtains may be glazed, depending on compatibility with expected smoke temperatures.

Design issues
- The volume of the smoke reservoir and the smoke extraction rate must provide an adequate ASET.
- The smoke extract rate should also prevent the build-up of smoke from overflowing the reservoir.
- If the reservoir is too large, the smoke may cool by losing heat to the structure. The smoke may lose buoyancy, and fall down to low levels where people could be endangered.
- Materials used within the smoke reservoir (including automatic smoke curtains) should be able to tolerate the predicted smoke layer temperatures.
- Sprinklers may need to be installed in the smoke reservoir if:
  - ❏ the layer temperature is high enough to endanger occupants due to thermal radiation from the smoke layer (typically > 200 °C).
  - ❏ there are sufficient combustibles under the reservoir to give a significant threat of excessive fire spread.
- Screens and curtains should channel the smoke so that it does not present a hazard to occupants on floors between the levels of the fire and the reservoir.
- Mechanical extract systems relying on a secondary power source (eg. a generator) will have a limited duration of operation, restricting the ASET that can be achieved.

Enforcement issues
- As for design

References
- *Design methodologies for smoke and heat exhaust ventilation* (BR 368), chapter 6
- *The SFPE handbook of fire protection engineering*, chapter 3-9

Links to other topics
SHEVS [2-11], Available Safe Egress Time [2-2], Smoke curtains [2-9]

Smoke curtains

Input ◄──────► Output

Description

When designing a smoke and heat exhaust ventilation system (SHEVS) it is often desirable to include features such as channelling screens or to block openings that would not be feasible for the normal day-to-day use of the building. A solution is to provide curtains that are activated by the smoke control system. These curtains must be designed to operate reliably and safely (occupants should not be in danger of being struck by an operating curtain). They should also be able to withstand the anticipated smoke temperatures. Free-hanging curtains can be deflected (and not provide an effective seal) due to the buoyant pressures of the hot gases contained to one side of the barrier, or by air movement caused by the extraction system.

Design issues

● The operation of the curtains should be 'fail safe', or else should be highly reliable.
● The curtain material should tolerate the design temperatures of the smoke layer (typically < 200 °C except in the case of channelling screens where the temperature may be 600 °C or higher).
● Operation of the curtains should not endanger building occupants.

Enforcement issues

● As design

References

● *Design methodologies for smoke and heat exhaust ventilation* (BR 368), chapter 6
● *BS 7974: Application of fire safety engineering principles to the design of buildings*. Part 2: Spread of smoke and toxic gases within and beyond the enclosure of origin, section 9.10

Links to other topics

SHEVS [2-11]

Further notes

CEN terminology is now 'smoke barriers' rather than 'smoke curtains'.

User's notes

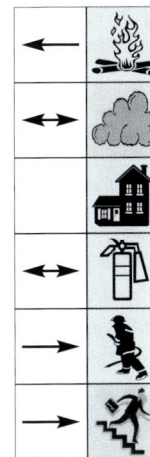

Smoke venting

Description

Smoke can be removed from a building either by natural or mechanical ventilation. Natural ventilation uses openings in the roof or side of the building for smoke to leave using its own buoyancy. Several mechanisms are available to open vents. These include servos activated by the smoke detection system, or fusible links. Mechanical systems use fans and (usually) a network of ducts to extract the smoke. Some fans may be directly mounted on the roof. The smoke is usually extracted from a reservoir created at roof level in the building. However, extraction using slit vents across openings (eg across the opening of a shop into a shopping mall) is also used.

User's notes

Design issues

● Effects of wind on natural ventilation systems should be considered.
● Smoke escaping from ventilation outlets should not be able to re-enter the building or affect other buildings.
● Provision of replacement (makeup) air should be considered.
● Fans and ductwork may need to operate at high temperatures.
● Excessive localised extraction can draw air from low level through the smoke layer creating a 'plug-hole' effect and reducing the efficiency of the extraction system. This is a particular problem if the smoke layer is shallow.

Enforcement issues

● A ventilation system cannot be designed in isolation from other parts of a building's fire protection system. The selection of design fire is critical. Fire resistance of ductwork should be considered.
● In the UK, fusible links tend to be regarded as a last-chance backup when primary opening devices have failed.

References

● *Design methodologies for smoke and heat exhaust ventilation* (BR 368), chapter 6
● *The SFPE handbook of fire protection engineering*, chapter 4-12
● *BS 5588 Fire precautions in the design, construction and use of buildings*. Part 7: Code of practice for the incorporation of atria in buildings

Links to other topics

SHEVS [2-11], Available Safe Egress Time [2-2], Replacement air [2-12]

SHEVS (Smoke and Heat Exhaust Ventilation Systems)

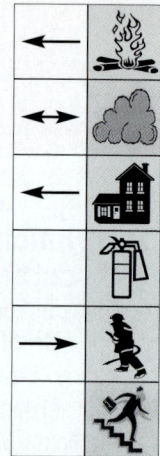

Input ◄──► Output

Description

Smoke and heat exhaust ventilation systems (referred to by the acronym, SHEVS) are designed to provide control of smoke from a given design fire. They are usually employed in large building spaces (eg shopping malls and exhibition halls) where there is no internal compartmentation and travel distances are long. The principles can, however, be applied to other building spaces where smoke control is required.

The principles of SHEVS are simple: hot buoyant gases from a fire rise to form a stable layer in a reservoir below the ceiling such that a clear layer of sufficient height may be present for long enough to achieve the safe evacuation of occupants and/or rapid firefighter access to the fire with good visibility. Smoke extraction from the layer may be by natural buoyancy or fans.

Design issues
● Correct assessment of the design fire
● Use of sprinkler and detection systems
● Escape times
● Compartmentation, downstands, smoke curtains
● Sizing of extract system (plume entrainment, etc.)
● Supply of replacement (makeup) air
● Building management system

Enforcement issues
● Design of SHEVS is more than calculations to determine the correct fan capacities: the interaction of the system with the overall fire safety strategy needs to be considered.
● In complex or unorthodox cases, or where safety factors appear marginal, verification of the design calculations using CFD simulations or hot smoke tests may be required.

References and resources
● *Design methodologies for smoke and heat exhaust ventilation* (BR 368)
● *The SFPE handbook of fire protection engineering*, chapters 4-12, 4-13
● *CIBSE guide E: Fire engineering*, chapter 7
● CFD and zone models

Links to other topics
This topic has strong links to many other topics notably:
Design fire [1-13], Fire plume [1-2], Spill plumes [2-4], Smoke venting [2-10], Required Safe Egress Time [6-1], Available Safe Egress Time [2-2], Smoke reservoirs [2-8]

2 12 Replacement air (or Makeup air or Inlet air)

Input ◄───► Output

Description

User's notes

When a smoke extraction system (either natural or powered) in a building is running, the mass of smoke extracted must be replaced by an equivalent mass of replacement or 'makeup' air (otherwise the smoke control system would attempt to create a vacuum in the building!). Restricting the flow of replacement air will make the smoke control system less efficient.

Replacement air will usually enter the building at low level where high air velocities may cause a number of problems. If the replacement air is drawn through an escape route, high velocities may impede the egress of occupants.

In naturally ventilated systems, where there may not be sufficient inlet area at low level, it may be possible to use open vents in an adjacent reservoir to provide sufficient replacement air.

Design issues

- A smoke extraction system will not function correctly without an adequate supply of replacement air.
- Beware of locating air inlets so that they feed directly into the extraction system causing a 'short-circuit'.
- The air inlets should be remote from the exhaust of the smoke extraction system so smoke is not 'recycled' through the building.
- Doors that open automatically, as part of the SHEVS, should be sequenced so that they are fully open by the time the extract fans are running at full capacity.

Enforcement issues

- The provision of and routes for replacement (makeup) air should be confirmed.
- If the occupants of the building are expected to use the route for replacement air then the velocity of airflow should be less than 3 m/s.

References

- *Design methodologies for smoke and heat exhaust ventilation* (BR 368), chapters 2 and 6
- *The SFPE handbook of fire protection engineering*, chapter 4-12

Links to other topics

SHEVS [2-11], Smoke venting [2-10]

Further notes

Replacement air will not 'feed the fire', it will tend to reduce the likelihood of an underventilated fire (giving toxic combustion products and a possibility of a backdraught) and provide clear air at low level, facilitating firefighter access and occupant egress.

A British Standard (*BS 7346*: Part 4) on steady-state SHEVS design is due for imminent publication.

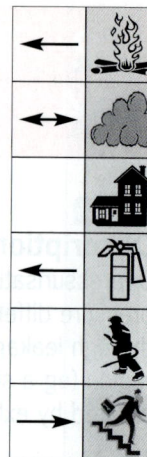

Hot smoke tests

Description

The ultimate test of any smoke control system is its performance in a real fire. A hot smoke test is based on a controlled fire designed so that the performance of the smoke control system can be monitored with no (or minimal) damage to the building structure. This provides a test of the whole system.

Methods developed in Australia, the UK and Belgium use clean-burning alcohol fires and artificial smoke to visualise the hot gases.

Cold smoke testing will not create the buoyant layers and rising plumes required to provide a realistic representation of smoke movement from a real fire.

Design issues

● Hot smoke tests provide a means of demonstrating to enforcers the performance of an installed smoke control system.

Enforcement issues

● A hot smoke test may be desirable when:
  ❏ there is doubt in the validity of assumptions used in the design of the smoke control systems,
  ❏ when the final installation is found to differ from the proposal used to gain approval,
  ❏ where conditional approval has been granted subject to proof of performance of the smoke control system.

The relationship between the size of fire used to design the system and the size of fire used for testing should be clear.

References and resources

● *Design methodologies for smoke and heat exhaust ventilation* (BR 368), chapter 15
● Equipment described in BR 368

Links to other topics

SHEVS [2-11], Smoke venting [2-10]

3 Structural fire protection

Also termed 'Fire resistance' [section 3-1], structural fire protection is an important facet of any fire safety design. It is needed to prevent the spread of fire and structural collapse, and is of increasing significance as a fire continues to grow to full room involvement and beyond.

Fire resistance is important in the protection of property, whether this is the building itself or its contents. Additionally, fire resistance has an impact on life safety, particularly in large or tall buildings, or where the occupants cannot move, move only slowly or may be asleep. In these cases, it must provide sufficient time for evacuation, or for people to remain in place while the fire is attacked by suppression systems or the Fire Service. It is also necessary to protect the members of the Fire Service, who may remain in the building well after all the other occupants have left. Additionally, fire resistance is important in the protection of property, whether this is the building itself or its contents.

Prevention of fire spread within the building is accomplished by sub-dividing the building interior into compartments. The compartment boundaries (walls, floors and ceilings) need to inhibit the spread of fire (and smoke, although the tests in *BS 476* do not examine this directly) to adjoining compartments, at least for a time, and thus slow down the spread of fire. Fire spread to nearby buildings could arise as a result of flame contact [section 3-8] or radiation heat transfer [section 3-9], either through windows or other openings, or openings that arise from structural collapse. The boundary walls therefore need to remain intact for as long as possible, and any windows limited in size to reduce radiant heat to adjacent buildings. Prevention of collapse requires loadbearing elements of the structure to retain stability for the duration of the fire.

Fire resistance is most commonly expressed in terms of time [section 3-1]. In furnace tests, the fire resistance time is the duration for which the structural element retains its stability/insulation/integrity, in the test according to the pass/fail criteria. However, in furnace tests it is important to note that:

- the heating regime may be very different to that experienced in a real fire, and
- furnaces can only test elements of a structure, not the entire structure.

Various calculation methods may be employed to predict fire resistance. It is important to realise that the time rating from a fire resistance test is a comparative measure, and is not the survival time in a real fire.

All calculation methods start with a definition or calculation of the heating regime. This enables calculation of the heat transfer from the compartment atmosphere to the structure [section 3-7]. The temperature (distribution) within the structure can therefore be evaluated, and this is used either as input to structural stability calculations, or directly to estimate the insulation resistance criteria. *Note:* calculations to predict loss of integrity (eg the opening of small fissures through which hot fire gases can pass and ignite material on the other side of the barrier) are currently not possible.

The fire severity [section 3-2] is defined in various ways as a measure of the destructive potential of the fire. However, the commonest and most convenient way is to express it in terms of time [sections 3-3, 3-4], ie an equivalent exposure to the heating regime of a furnace test, that causes a similar level of damage. It is then possible to compare fire resistance and severity in terms of a common unit (time), with the design objective being that the resistance exceeds the severity by an appropriate safety margin [section 7-5]. Since both the severity and the resistance are random variables, there will always be some probability of failure (ie untimely collapse). The design objective is therefore to keep this acceptably small (see *Safety factors* [section 7-5]).

The fire severity depends on the temperature reached by the fire within the compartment, and also the duration of the fire. Both of these quantities in turn depend on the amount of fuel available (the 'fire load' [section 3-5]), and its physical disposition (eg whether in thin wall linings, or solid blocks of material).

User's notes

3 Structural fire protection

Effect **of** structural behaviour	Effect **on** structural behaviour	
Fire growth ● If integrity or insulation failures occur, fire will spread into adjoining compartments	**Fire growth** ● Localised heating of structural elements may occur, for example where the fire plume impinges on the ceiling ● Fully involved compartment fires transfer heat to all structural elements comprising the compartment boundary ● Structural loadbearing resistance, integrity or insulation failures may occur as a consequence of heat transfer	
Smoke spread ● Compartmentation can restrict the movement of smoke around the building ● If integrity failures occur, smoke will spread into adjoining compartments	**Smoke spread** ● Smoke layers will transfer heat to the structure ● Preheating of adjoining compartments may make their failure more rapid when fire spreads to them	
Detection/Suppression ● Compartmentation may concentrate smoke, and lead to earlier detection ● Conversely, compartmentation may prevent smoke from getting to a detector ● Some suppression systems only work effectively in closed compartments	**Detection/Suppression** ● Suppression systems will usually extinguish, or at least prevent further growth of, the fire once they are activated, and hence reduce the heat transfer to the structure ● Thermal shock, due to sudden cooling by sprinkler sprays, may cause windows to break	
Fire Service intervention ● Structural collapse is a hazard for personnel inside the building	**Fire Service intervention** ● Reduce heat transfer to structure, as for automatic suppression ● Breaking windows or creating openings in the roof can remove smoke	
Human factors ● Structural collapse, while less likely to be a hazard for people (other than Fire Service) inside the building, may nevertheless be a hazard to those outside	**Human factors** ● Leaving fire doors open may allow the spread of fire	

User's notes

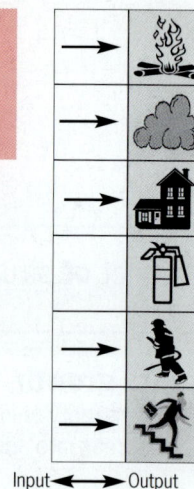

Input ←——→ Output

Description

The term 'fire resistance' can either be synonymous with 'passive fire protection' (ie construction materials and coatings which enable the building to withstand the fire), or else relates to a period of time for which an element of construction (beam, column, floor, wall, etc.) will survive in a standard fire test. There are three performance criteria:

- **stability** (avoidance of structural collapse or unacceptable deformation),
- **integrity** (avoidance of cracks and fissures),
- **insulation** (restriction on temperature of unexposed face).

The corresponding terms in the Eurocodes are **loadbearing function (R)**, **integrity (E)** and **insulation (I)**.

User's notes

Design issues

- Fire resistance time is to the first failure criterion, although other failure modes may occur soon afterwards
- The 'standard fire curve' [3-3] is not a good model for a real fire; the latter may have a faster growth rate and higher peak temperature, but a finite duration.
- Localised heating may occur in real fires, but not in standard fire tests where furnace temperatures are considered to be uniform.
- The performance of the structure as a whole may be quite different to that of individual components, due to redistribution of loads.

Enforcement issues

- There is no direct relationship between performance in a standard test and the duration of a real fire. It is a myth that the 'fire resistance time' is the period that a structure will survive in a real fire.

References

- *Structural design for fire safety*, chapter 6
- *Building and fire*, chapter 8
- *CIBSE guide E*, chapter 5
- *The SFPE handbook of fire protection engineering*, chapters 4-9, 4-10, 4-11

Links to other topics

Fire severity [3-2], Time-temperature curves [3-3], Structural modelling [3-7]

Further notes

Typical performance criteria are listed below.
- **Stability:**
  - ❏ Beams — deflection < length/30
  - ❏ Columns — a failure to support the applied load
- **Integrity:** ignition of a cotton pad held close to an opening
- **Insulation:** a temperature in excess of 140 °C (average) or 180 °C (single point)

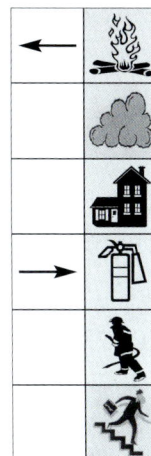

Description

Fire severity is a measure of the destructive impact of a fire, or a measure of the forces or temperature that could cause collapse or other failure as a result of a fire.

Design issues

- The severity may depend on the total amount of heat transferred to the structure, or the peak temperature attained.
- Severity may be expressed in terms of:
  - ❏ time (equivalent exposure to a 'standard' fire curve in a furnace),
  - ❏ temperature (maximum reached by key part of structure), or
  - ❏ minimum loadbearing capacity.
- Real fires have very different time-temperature curves to the standard curve so expressing severity in terms of equivalent exposure may be difficult. It may be necessary to perform calculations from 'first principles'.

Enforcement issues

- Fire severity must be compared with the fire resistance of the structure (in terms of time, temperature or loadbearing capacity) and an adequate safety margin must be demonstrated.
- Attempts to express severity in terms of equivalent exposure to a 'standard' fire (so that the 'fire resistance' can be taken directly from fire test results) may not be valid for some structures.

References

- *Structural design for fire safety*, chapter 5
- *Building and fire*, chapter 5
- *CIBSE guide E*, chapter 5

Links to other topics

Time–temperature curves [3-3], Time-equivalent exposure [3-6]

Further notes

In post-flashover fires, heat transfer is primarily by radiation, the rate of which is proportional to the 4th power of absolute temperature. (As both the hot gases and the structure will be radiating heat, the net transfer is proportional to $\varepsilon_{gas}T^4_{gas} - \varepsilon_{struct}T^4_{struct}$). Real fires, which usually attain higher temperatures than the 'standard' fire in a fire test, but have shorter duration, and will therefore have quite different heat transfer to the standard fire.

User's notes

Input ◄──► Output

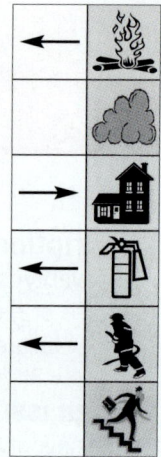

Input ◄───► Output

Description

Time–temperature curves are used to describe conditions within a fire compartment, for post-flashover or fully involved fires. The determination of the fire resistance of structural elements is based on the exposure of a test specimen in a furnace whose temperature follows a standard curve.

User's notes

Design issues

● Time–temperature curves are included in calculations of fire resistance.
● Real fires may be quite different from the standard curves – their severity may need to be expressed in terms of an equivalent time of exposure to a standard curve.

Enforcement issues

● Fire resistance time is not related to the time available for escape, in most situations.
● Standard curves take no account of fire load density, compartment size, or ventilation.
● Time–temperature curves are a feature of standard test methods and do not necessarily provide data that can be used in a performance-based design.

References and resources

● The SFPE handbook of fire protection engineering, chapter 4-8
● An introduction to fire dynamics, chapter 10
● Fire safety engineering design of structures, chapter 3
● Computer models, eg COMPF2, OZONE

Links to other topics

Fire severity [3-2], Fire resistance [3-1]

Further notes

The equation for the ISO standard curve, intended for rooms with mainly wood-based contents is:

$$T = 20 + 345 \log_{10}(8t + 1)$$

An alternative curve can be used for hydrocarbon or plastic contents:

$$T = 1100(1 - 0.325e^{-0.1667t} - 0.204e^{-1.471} - 0.47e^{-15.833t})$$

In both of these equations, T is in Celsius and t is in minutes.

The 'standard' Time–temperature curves do not represent a specific fire scenario, consequently the period of fire resistance from a furnace test does not necessarily relate to the time that the building element will survive in a real fire.

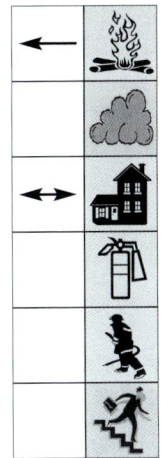

Parametric fire Time–temperature curves

Description
Unlike the 'standard' Time-temperature curves [3-3] where the compartment fire temperature is a function of time and nothing else, parametric curves attempt to give a more realistic prediction of temperature dependent on a small number of key variables.

The key variables are typically:
● *fire load* (see section 3-5): the amount of fuel determining the duration of the fire,
● *ventilation conditions* affecting combustion efficiency and also rate of heat loss by convection,
● *compartment size*,
● *material properties* of the compartment walls determining the heat transfer from hot gases to the solid walls.

Design issues
● The parametric T(t) curve can be used as input to computer models predicting the heat transfer to the structure, and in turn for predictions of structural stability.

Enforcement issues
● Parametric curves have only been validated for compartments of moderate size (~100 m^2 floor area, 4 m ceiling height), a limited range of ventilation conditions, and fire loads of predominantly cellulosic fuels (rather than polymeric).

References
● *Structural design for fire safety*, chapter 5
● *Fire safety engineering design of structures*, chapter 4

Links to other topics
Fire resistance [3-1], Fire load [3-5], Fire severity [3-2], Time-equivalent exposure [3-6]

Further notes
Some examples of parameteric fire curves are shown below, along with the standard and 'hydrocarbon' curves.

Time-temperature graphs

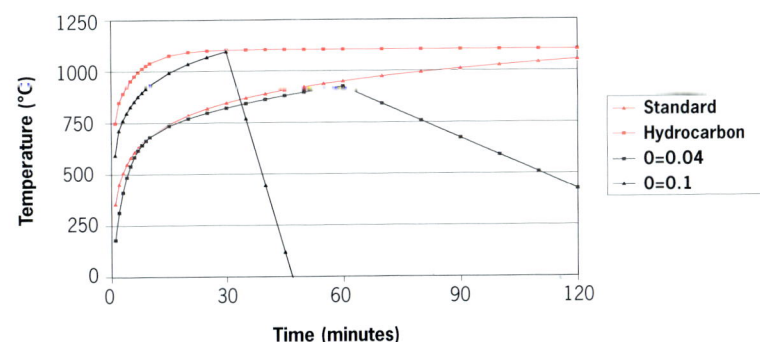

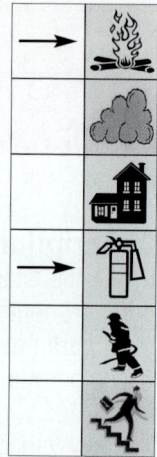

Input ←—→ Output

Description

Fire load is defined as the calorific value of the mass of fuel within a compartment. (It may also be expressed in terms of kg of wood, the calorific value of which may be taken as 16~18 MJ/kg). Fire load density is the fire load per m² of the compartment floor area, and is often assumed to be uniform for compartments of different size but similar function.

User's notes

Design issues

● Fire load is a key parameter in determining the duration of a fire, and hence its severity.

● Fire load is a probabilistic variable. It is common to take the 80% fractile value of the distribution (ie the value that is not exceeded in 80% of buildings of a given type) as the design load.

● If the load is expressed in terms of kg(wood)/m², there may be confusion between the full calorific value of 16~18 MJ/kg (as measured in a bomb calorimeter, not representative of real fire conditions), and the free-burn value of 13~14 MJ/kg (due to incomplete combustion even with unlimited air supply). It is better to express fire load as MJ/m²

● Fire load alone does not determine the HRR; the nature of the material (eg thin wall linings, solid blocks) will also be important.

Enforcement issues

● Parametric curves have only been validated for fire loads of predominantly cellulosic fuels (rather than polymeric), among other limitations [section 6-4].

References

● *Structural design for fire safety*, chapter 5
● *BS 7974 PD1: Initiation and development of fire*

Links to other topics

Fire resistance [3-1], Fire severity [3-2], Time-equivalent exposure [3-6]

Further notes

Some values given in BS 7974 (PD1) for fire load density are tabulated below. These are for guidance, and should not be regarded as prescriptive. For specific buildings, different values may be appropriate.

Fire load density guide values for different categories of building use	
Occupancy/ activity	Fire load density (80% fractile)
Dwelling	870 MJ.m⁻²
Hospital	350 MJ.m⁻²
Hotel room	400 MJ.m⁻²
Library	2250 MJ.m⁻²
Office	570 MJ.m⁻²
School	360 MJ.m⁻²
Shops	900 MJ.m⁻²

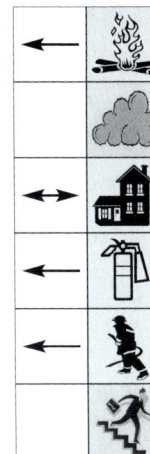

Time-equivalent exposure

Description
The fire severity of a real fire is expressed in terms of an equivalent exposure to a 'standard' fire test. (This is so that the fire resistance of the structural element can be taken directly from the test measurement).

'Time-equivalence' formulae have been derived empirically from calculations of the time for protected steel members to reach a failure temperature, exposed to realistic fires (using simple formulae for their T(t) history) and also to the 'standard' fire curve.

Design issues
● Use of time-equivalence formulae is generally accepted for protected steel members (for which the formulae were derived) and concrete, but not other types of structures (eg unprotected steelwork).

Enforcement issues
● Time-equivalent formulae are empirical, but the limitations are not well-documented.
● Determining equivalence on the basis of equal areas under the Temperature–time curves for the real fire (0 < t < infinity) and standard curve (0 < t < t_{equiv}) does not adequately reflect the heat transfer mechanism (radiation, proportional to T^4).
● Time-equivalent formulae may not be applicable to fire scenarios other than assumed in the original empirical derivation (eg non-cellulosic fuels, larger rooms, different types of fire protection, different levels of glazing/ventilation) or different types of structural members.
● Time-equivalent formulae are not intended for unprotected steel or timber structures.

References
● *Structural design for fire safety*, chapter 5
● *Introduction to fire dynamics*, chapter 10

Links to other topics
Time–temperature curves [3-3], Fire severity [3-2]

Further notes
Time-equivalent formulae are a crude approximation, attempting to introduce real fire behaviour into fire engineering calculations. Much more accurate results are obtained if calculations are made from first principles, starting from realistic T(t) curves.

User's notes

Input ◄─────► Output

Structural modelling

Input ◄────► Output

Description

Numerical modelling of the response of structures to a fire falls into two distinct stages. Firstly, thermal analysis calculates the heat transfer to the structure, on the basis of a (transient) Time–temperature curve for the fire environment. Then the mechanical response of the heated structure/members is calculated, and the interaction with the rest of the structure if appropriate.

User's notes

Design issues

- Most models have a similar theoretical basis, however differences arise due to the way in which material properties are modelled, and the data used.
- In thermal models, some boundary conditions (eg surface emissivity for heat re-radiation) tend to be adjusted empirically to improve the fit with experimental measurements.
- Thermal analysis of concrete structures may have significant errors if there is an inadequate treatment of moisture within the concrete.
- Structural analyses are very sensitive to the temperature of the structure, so it is essential that the thermal model predictions are accurate to begin with.
- Some of the simplifications adopted by structural models are inadequate, when modelling concrete structures.

Enforcement issues

- Modelling limitations are discussed above.

References and resources

- **Sullivan P J C, Terro M J & Morris W A.** Critical review of fire-dedicated thermal and structural computer programs. *J Applied Fire Science* 1994: **3**: 113–135.
- *The SFPE handbook of fire protection engineering*, chapter 4-9
- Computer software, eg THELMA, TASEF, CEFICOSS, LENAS, SAFIR

Links to other topics

Fire severity [3-2], Fire resistance [3-1], Time–temperature curves [3-3]

Further notes

Integrity cannot be modelled with the current state of the art methods.

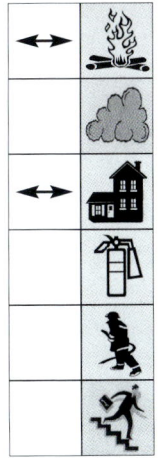

Flames from windows

Description

In fully involved room fires, window glazing may often fall out allowing flames to appear out of the windows. This can have important consequences for flame spread up the external building façade, causing fire spread by re-entry through windows higher up, the stability of external structural members, and radiation impact on adjoining buildings.

Input ←→ Output

Design issues

● The equations for the height and projection of external flames have been derived empirically, and are only approximate due to large scatter in the data.

Enforcement issues

● The empirical equations break down if:
  ❏ there are substantial heat losses to the façade,
  ❏ external winds deflect the flame and reduce its length,
  ❏ flames merge from more than one floor,
  ❏ the burning rate is greater than expected (eg the fuel has a large surface area, or the fuel is non-cellulosic, with a higher volatility).

References

● *An introduction to fire dynamics*, chapter 10
● *CIBSE guide E*, chapter 9

Links to other topics

Radiation from windows [3-9], Fire severity [3-2]

Further notes

The empirical equations (from chapter 10 in *An introduction to fire dynamics*, alternative expressions are also available in the literature) for flame geometry are:

$$z = H = 12.8(\dot{m}/B)^{2/3} \quad x/H = 0.454/(2B/H)^{0.53}$$

where H and B are the height and width of the window, respectively (m), $\dot{m}$ is the burning rate (kg.s^{-1}), z is the height of the flame tip above the soffit (m) and x is the projection away from the wall (m).

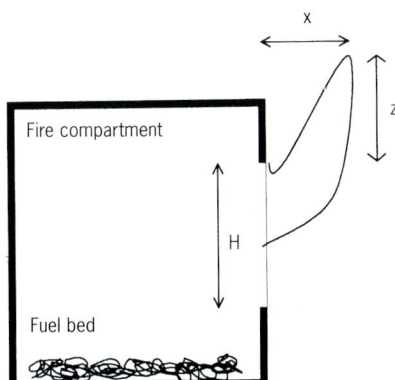

Input ⬅➡ Output

Description

To prevent fire spread from building to building, the effect of thermal radiation from a burning building to its neighbour must be considered. The calculation methods in BR187, referred to in Approved Document B, assumes that all the windows (unless insulated glass) in a compartment containing a fire can be considered to be radiating at a temperature determined by the fire load in the compartment. Tables and calculation aids are provided so that the **view factor** (**configuration factor**) and the incident heat flux on an adjacent building can be calculated. If the incident flux is below the level required for pilot ignition of wood (12.6 kW/m^2) then the distance between the buildings is satisfactory. In practice, the distance to the site boundary is calculated and it is assumed that the two buildings are identical.

The calculation methods in BR187 do not include the contribution of flames extending beyond an opening and the effects of wind.

Design issues

● Calculation methods are detailed and time-consuming but not fundamentally difficult.

Enforcement issues

● The calculation methods include a number of caveats to reduce the number of trivial calculations. There may be a temptation to exploit these to gain approval for some designs, particularly where there are small windows or doors on a wall close to a boundary.

References and resources

● *An introduction to fire dynamics*, chapter 2
● **Read R E H.** *External fire spread: building separation and boundary distances*. BR 187. Garston, BRE, 1991
● TenRad computer software (Tenos Ltd)

Links to other topics

Compartment fire modelling [1-9], Flashover [1-5], Flames from windows [3-8]

User's notes

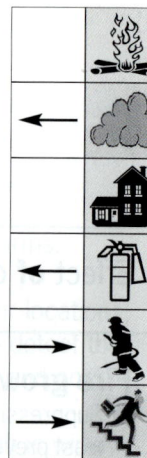

Smoke detection

Description
Smoke detectors usually work by sensing the presence of air-borne particulates but they may also work by chemical means (eg CO detectors). The presence of particulates, other than from fire (eg dust, water droplets from condensed steam), can lead to false alarms.

Ionisation detectors respond with a change in electrical current through an ionisation chamber in the presence of smoke particles. This is proportional to the number and size of the particles. They are best used to detect flaming fires.

Optical detectors may measure the attenuation of a light beam that is interrupted by smoke or light scattered by smoke particles in a chamber. They are best used to detect smouldering-type fires.

The sensor unit may be located at the detection point in an integrated unit (that may include a sounder) or remotely in an aspirated system where air is drawn through pipes from the sensing point(s) to a detector unit.

Optical and ionisation detectors respond differently to smoke from different fuel sources. It may not be possible to predict the response of a detector to the properties of smoke measured in tests, such as the cone calorimeter, because the detector may respond to properties measured under different conditions (eg wavelength of light source).

Design issues
- Optical beam detectors can be used in large spaces (eg atriums).
- Response will vary with different fuel sources which have different smoke properties.
- CFD simulation may be required to determine the optimum detector locations.

Enforcement issues
- As design

References
- *The SFPE handbook of fire protection engineering*, chapter 4-1
- *Buildings and fire*, chapter 10
- *CIBSE guide E*, chapter 6

Links to other topics
Smoke properties [1-3]

Further notes
A 'rule of thumb' is that ionisation detectors operate when smoke optical density (OD) is approximately 0.1 m^{-1}. As this is also often regarded as a tenability limit for escape routes, it follows that detectors must be sited to pick up smoke flows before the layer depth is such that it affects the escape route.

User's notes

Input ◄──► Output

4
2

Heat detection

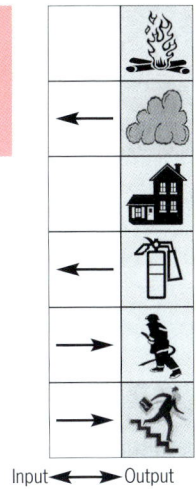

Description

Heat detectors sense temperature changes at the sensor location and are activated if some criterion (temperature or rate of rise of temperature) exceeds a threshold value that has been selected to discriminate between fire and non-fire conditions.

Sprinkler heads act as heat detectors as the bulb or fusible link breaks in response to the local gas temperature.

Design issues
● Response time is characterised by the Response Time Index (RTI).

Enforcement issues
● As design

References and resources
● *The SFPE handbook of fire protection engineering*, chapter 4-1
● *Buildings and fire*, chapter 10
● *CIBSE guide E*, chapter 6
● Computer models such as DETACT

Links to other topics
Response Time Index [4-4], Smoke detection [4-1], Detector location [4-5], Ceiling jets [2-3]

User's notes

Radiation detectors

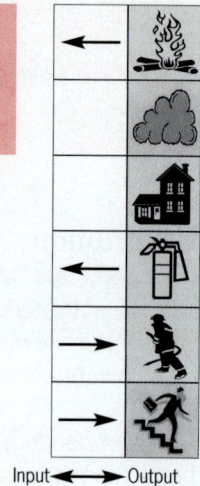

Input ◄────► Output

Description

The combustion process creates electromagnetic radiation over a wide spectrum and detectors are available that work in the ultra-violet (100–350 nm), visible (250–750 nm) and infrared (750–2200 nm) ranges. The selection of a particular sensor for a scenario depends on the nature of the fuel and the background environment.

Radiation sensors are line-of-sight devices and must be able to 'see' the potential fire source, but be shielded from potential erroneous sources such as the sun.

Radiation detectors can be made to respond very rapidly so that they can be used in explosion suppression systems.

User's notes

Design issues

● The sensor must be matched to hazard and normal environment.

Enforcement issues

● As design

References

● *The SFPE handbook of fire protection engineering*, chapter 4-1
● *Buildings and fire*, chapter 10
● *CIBSE guide E*, chapter 6

Links to other topics

Smoke detection [4-1], Heat detection [4-2]

Further notes

1 nm (nanometre) = 10^{-9} m

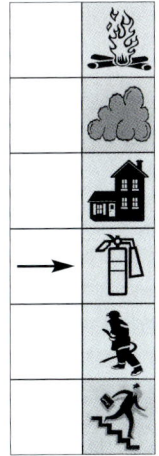

Response Time Index (RTI)

Description
The Response Time Index (RTI) is a measure of how rapidly a heat detector or sprinkler head will react to a rise in temperature. The other main factor influencing the response time is the nominal operating temperature (of the detector, not the local gas temperature).

RTI is constant for any given device. A small value indicates a faster response.

Typical values for sprinklers	
Sprinkler type	**RTI [$m^{1/2}s^{1/2}$]**
Standard	80–200
Special response	50–80
Quick response	50 or less

Design issues
● Used in calculations of detector/sprinkler activation time

Enforcement issues
● The device does not react as soon as the hot gas temperature equals the nominal operating temperature.
● Factors such as heat conduction to the sprinkler pipes, sprinkler orientation and airflow deflection, and the latent heat of fusion for solder links, may introduce delays in the sprinkler activation time.

References and resources
● *The SFPE handbook of fire protection engineering*, chapter 4-1
● *Buildings and fire*, chapter 10
● *CIBSE guide E*, chapter 8
● Computer models such as DETACT

Links to other topics
Ceiling jets [2-3], Detector location [4-5]

Further notes
The equation describing the rise in temperature of the detector or sprinkler head is:

$$\frac{dT_d}{dt} = \frac{u^{0.5}(T_g - T_d)}{RTI}$$

where T_d is the device temperature, T_g is the hot gas temperature (both in units of Kelvin), u is the velocity of the hot gas in the ceiling jet flowing past the device (ms^{-1}). All these variables are time-dependent, so the equation must be integrated to find T_d at any time. RTI is constant, with units of $m^{0.5}s^{0.5}$.

When T_d reaches the operating temperature, the device activates.

Detector location

Input ◄────► Output

Description

The best possible location for a detector would be immediately above the fire. However, except in scenarios with a clearly identifiable hazard, the fire location will be unknown and a number of detectors will be required to cover the whole building. This is usually achieved with a grid of detectors at ceiling level. Ceiling jet calculations can be used to estimate the activation times of thermal detectors. Calculations for optical and ionisation detectors are more uncertain because of the difficulty in predicting the smoke properties that are sensed by the detector.

Radiation detectors need to be located so that their field of view is not obstructed. This will usually require the use of several detectors to cover a particular area.

Beam detectors need to be located considering the behaviour of the associated smoke control system so that layers and rising plumes in large spaces can be detected.

User's notes

Design issues

● When a fire starts the HRR is usually small and the normal air movement in the building will dominate the initial movement of smoke and hot gases. It is during this phase that detection of the fire is most desirable.
● In the case of complex geometries, or where there is complex pre-fire air flow, the use of CFD models to examine the response of detectors may be useful.

Enforcement issues

● As design

References

● *The SFPE handbook of fire protection engineering*, chapter 4-1
● *Buildings and fire*, chapter 10
● *BS 5839: 2002: Fire detection and alarm systems for buildings.* Part 1: Code of practice for system design, installation, commissioning and maintenance

Links to other topics

Buoyancy [2-1], Ceiling jets [2-3], Fire plume [1-2]

Further notes

In some applications, there may be a trade-off between siting detectors for the minimum possible response time and the possibility of false alarms. For example, a domestic detector located in a kitchen would ensure rapid detection of a chip pan fire but would activate frequently during cooking.

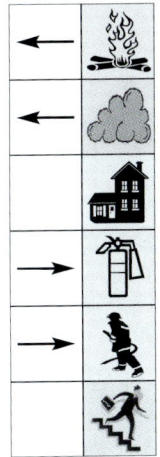

Sprinklers

Description

Automatic sprinklers and the rules for their installation have been available since the 1880s and are now widely used in commercial applications and, increasingly, in residential premises. A sprinkler system is a system of pipework to distribute water to spray heads in the event of a fire. In an automatic system, the spray heads include a temperature sensor, usually a glass bulb or fusible metal link, that will allow the flow of water if a critical temperature is exceeded.

The most common systems are 'wet' (the pipework contains water under pressure up to the sprinkler head), however 'dry' systems are used where the water in the pipework could freeze. Deluge systems do not have automatic heads but activate a supply of water to open heads in response to a detector system. There are also pre-action systems where the pipework is 'dry' until activation of a detector system, but the release of water is controlled by automatic heads.

In addition to providing a system to control or extinguish, automatic sprinklers also detect the fire.

Design issues
● Extensive prescriptive rules exist for the design of sprinkler pipework and associated water supplies.
● The addition of sprinklers often leads to relaxation of other (prescriptive) requirements, for example fire resistance times, and building separation distances may be reduced (Approved Document B, Building Regulations (England & Wales), Section 14.17).

Enforcement issues
● As design

References
● BS 5306: 1990: Fire extinguishing installations and equipment on premises. Part 2: Specification for sprinkler systems
● The SFPE handbook of fire protection engineering, chapter 4-3

Links to other topics
Response Time Index [4-4], Detector location [4-5]

Further notes
The response and action of sprinklers is being included in some fire simulation models. This should be regarded as a 'research feature' and the assumptions included in the sub-model considered carefully before any results are used in a fire engineering design.

User's notes

Input ◄───► Output

Other suppression systems

Input ◄─────► Output

Description

While water is an excellent extinguishing agent due to its high specific heat capacity, latent heat of vaporization and availability, it is not suitable for all types of fire. For example, water can react explosively with some materials (especially when they are at high temperatures). If we recall the fire triangle [G-1], a fire can be extinguished by removing one (or more) of the three components of heat, fuel and oxygen. Water acts by removing heat (and in some cases oxygen by displacing it with steam). Firefighting foams form a barrier to restrict the availability of oxygen (and also provide some cooling). Foams can be applied in fixed installations similar to sprinkler systems. Flooding with an inert gas (eg nitrogen) displaces oxygen.

Halons provide an efficient extinguishing action by inhibiting the combustion chemistry and stopping the chain reactions from creating heat. They are effective for many types of fire. However, their production was discontinued in 1994 as a result of the Montreal Protocol to protect stratospheric ozone. Alternatives are being developed.

The development of water mist systems is also current.

User's notes

Design issues

● Where a specific fire risk and fuel can be identified, a compatible extinguishing system can be selected.

Enforcement issues

● Where gaseous flooding suppression systems are specified, a room integrity test will demonstrate that the system will work efficiently as designed without leaking gas to other rooms or the outside.

References

● *The SFPE handbook of fire protection engineering*, chapters 4-4 to 4-7

Links to other topics

Fire triangle [G-1], Sprinklers [4-6]

5 Fire Service intervention

The design of a building should not assume that fire service intervention will contribute to the evacuation of occupants. Rescue by the Fire Service should be seen as an additional safety factor and not as an accepted part of the building evacuation. However, the first concern of the Fire Service is to ensure that all occupants have reached a place of safety before the main attack on the fire is made.

The time taken for the Fire Service to begin activity to control [section 5-3] and extinguish the fire depends on:
● the time of notification (related to detection time),
● the time taken to arrive at the scene of the fire [section 5-1], and
● the time to assemble the required resources [section 5-2].
As the firefighting activity proceeds, further resources may be required.

Adding the times for Fire Service activity to the overall time line for a fire scenario can give the size (Heat Release Rate) of the fire when firefighting activity begins. This will give an indication of the firefighting resources required and a means of examining the effect of different fire protection measures (eg different detection systems) on the conditions that the firefighters will encounter.

Times for Fire Service actions are difficult to quantify and calculation methods are not available. Qualitative judgement in consultation with the Fire Service is required.

The time taken for the Fire Service to begin effective firefighting operations depends on the ease of access to the seat of the fire. There must be space outside the building for the deployment of fire appliances and other equipment. In some cases (eg tall buildings or deep basements), it may be necessary to provide protected routes within the building (firefighting shafts, lifts, and corridors [section 5-5]). These will enable the firefighters to move through relatively smoke-free air until they get close to the fire.

Firefighting tactics such as positive-pressure ventilation [section 5-4] can be used to give the firefighters a relatively smoke-free route to the fire. However, a possible drawback is that other regions of the building may become smoke-logged.

User's notes

5 Fire Service intervention

Effect **of** Fire Service intervention	Effect **on** Fire Service intervention	
Fire growth ● Manual suppression will extinguish or control the fire growth	**Fire growth** ● Rapid increases in fire growth (eg flashover or backdraft) may be hazardous to personnel	
Smoke spread ● Manual suppression will extinguish or control the fire growth, and hence reduce smoke production ● Fine sprays of water into hot smoke layers can cool them down ● Windows may be broken or openings made in the roof to remove smoke	**Smoke spread** ● Reduction in visibility caused by smoke will hamper fire service operations within the building	
Structural behaviour ● Reduce heat transfer to structure, as for automatic suppression ● Windows may be broken or openings made in the roof to remove smoke ● Nearby structures will be protected	**Structural behaviour** ● Structural collapse is a hazard for personnel inside the building	
Detection/Suppression ● Manual suppression will extinguish or control the fire growth, complementing the effect of automatic systems	**Detection/Suppression** ● There will be no intervention until detection and warning has occurred ● Automatic detection, coupled with automatic warning of the Fire Service, may lead to a quicker response ● Detectors can tell the Fire Service where to find the fire ● Suppression, if it does not extinguish the fire, at least means the Fire Service has a smaller fire to tackle	
Human factors ● Arrival of the Fire Service may be the cue that persuades people the fire is 'real' ● The Fire Service can rescue people unable to escape by themselves	**Human factors** ● People may call the Fire Service to the fire	

User's notes

Arrival time

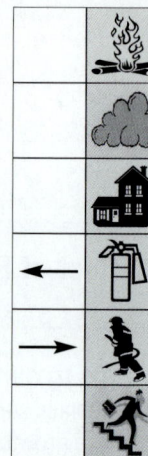

Input ◄──► Output

Description

The time taken for the Fire Service to arrive at a fire scene depends on:

● travelling distance,
● nature of route and traffic conditions,
● weather conditions,
● access to the fire site,
● information given about fire location on site,
● prior commitments of closest appliances.

Arrival time is measured after notification of the Fire Service. This is after the detection time of the fire.

Design issues

● In the UK, arrival time can be assumed to comply with standards of cover recommended by the UK government's Home Office. These range from 5 to 20 minutes depending on the risk category for the building.
● Advice should be sought from the Fire Service regarding the initial level of attendance for a building.
● It is important to provide sufficient hard-standing and fire hydrants close to the building, for a reasonable number of appliances.

Enforcement issues

● As design

References

● *BS 7974: 2001: Application of fire safety engineering principles to the design of buildings. Code of practice. PD5 Fire service intervention*

Links to other topics

Available Safe Egress Time [2-2], Required Safe Egress Time [6-1], Set-up time [5-2], Control time [5-3]

Further notes

Access requirements for the Fire Service are covered in Approved Document B5, section 17, of The Building Regulations (England & Wales).

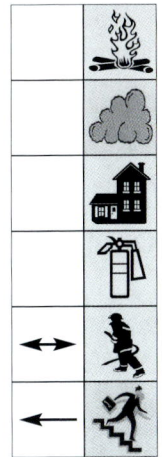

Input ◄──► Output

5 2 Set-up time

Description

Firefighters cannot begin to fight the fire the moment they arrive on site. On arrival, part of their effort will be concerned with aiding evacuation and rescue of the building occupants. A conservative assumption is that firefighting will only start after evacuation is complete.

Time will also be required to deploy any specialist equipment brought in from off-site or provided on the premises for fire service use.

User's notes

Design issues

● Firefighting usually starts after evacuation is complete.
● Consultation with the Fire Service is required to establish the set-up time for a particular fire scenario.

Enforcement issues

● As design

References

● *BS 7974: 2001: Application of fire safety engineering principles to the design of buildings. Code of practice.* PD5 Fire service intervention

Links to other topics

Available Safe Egress Time [2-2], Required Safe Egress Time [6-1], Arrival time [5-1], Control time [5-3]

Control time

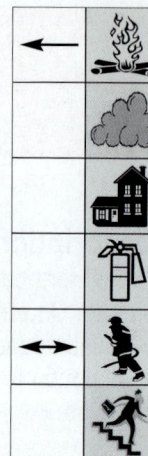

Description

Control time is the time taken for firefighters to stop a fire growing and start to make headway with extinguishing it. The time taken for firefighters to reach the point where they are confident that it is under control is dependent on a number of factors such as:

● fire size at detection,
● fire growth between detection and the start of firefighting,
● result of any first-aid firefighting,
● action of fire suppression systems (eg sprinklers),
● techniques used by firefighters,
● presence of protected access (eg SHEVS, pressurised stairwells).

UK fire reports do not indicate a maximum fire area that can be controlled by the first attendance.

References
● BS 7974: 2001: *Application of fire safety engineering principles to the design of buildings. Code of practice.* PD5 Fire service intervention

Links to other topics
Available Safe Egress Time [2-2], Required Safe Egress Time [6-1], Arrival time [5-1], Set-up time [5-2]

User's notes

Input ◄──► Output

5 / 4 Positive Pressure Ventilation (PPV)

Description

Positive pressure ventilation (PPV) is a technique used by firefighters to remove smoke from a building so that they can gain access for firefighting purposes. The technique involves placing a portable fan (usually carried on the fire appliance) so that it is facing an opening to a building and providing exhaust openings on the other side of the building or on a higher storey. Firefighters can then advance in clear air supplied by the fans.

Although an apparently simple technique, a detailed knowledge of the building is required to select openings to be used as exhaust vents. The effect of wind must also be considered otherwise the flow of fire gases may not be in the anticipated direction (pieces of foil can be used as flow direction indicators).

Design issues

- This is a firefighting tactic. Fans are carried on fire appliances.
- Firefighters need to be able to identify windows and doors in the building that can be used as exhaust vents. These will vary with fire location and entry point.

Enforcement issues

- As design
- The technique should not be used where it might push smoke so that it affects escape routes or occupants elsewhere in the building

References

- www.firetactics.com

Links to other topics

Effects of wind [2-6], Smoke venting [2-10]

User's notes

Input ◄──────► Output

Firefighting shafts, lifts and corridors

Input ◄──► Output

Description

In tall buildings, firefighters will require access to upper levels of the building. Similarly, in deep basements they will require access to the lower levels. This may be provided by protected shafts which contain stairs, and usually a lift, having a specification to enable use by firefighters during a fire.

In a deep building, protected corridors can fulfil the same function horizontally.

User's notes

Design issues

● Firefighting lifts require independent power supplies and controls that provide the firefighters with full control of the lift and doors overriding the normal call points on each floor.
● Firefighting lifts should not be used for general evacuation, as they should be waiting on the ground floor for use by firefighters on arrival.
● However, firefighting lifts may be used for the evacuation of disabled people.

Enforcement issues

● Annual testing may be required by the fire authority.

References

● *BS 5588: 1991: Fire precautions in the design, construction and use of buildings.* Part 5: Code of practice for firefighting stairs and lifts.
● *BS 5588: 1999: Fire precautions in the design, construction and use of buildings.* Part 8: Code of practice for means of escape for disabled people
● **Harrison R & Miles S.** *Smoke shafts protecting fire shafts — their performance and design.* BRE Report 79204 (CD-Rom). Garston, BRE Bookshop. 2002

Links to other topics

Effects of disabilities [6-10]

Further notes

Firefighting shafts are expected to serve as a means of escape as well as for Fire Service access.

6 Human factors

In any fire safety engineering design, protection of the building occupants is of paramount importance. The design will be heavily influenced by the initial distributions of people, which in turn will depend on the intended use of the building. The total number of people depends on the building's size, but their distribution within the building depends on the intended use of the different rooms. The time of day or night may also have a dramatic impact on the population. It is also important to consider not just numbers of people, but also other characteristics which will have an influence on how they behave, and hence the outcome of any fire scenario. People's behaviours may depend on many factors, for example a defined role within the building population, the influence of training, familiarity with the building, age, gender, and other factors such as disabilities [section 6-10] which may restrict the activities that may be attempted.

People's behaviour is firstly of significance in responding to the initial fire cues [section 6-2]. The information content of the cues is of greatest importance in determining people's reaction to them. An uninformative cue is likely to be ignored, until reinforced by a stronger cue. The person's current activity at the time of the first cue will also influence how likely they are to change this activity and respond to the cue. A person with a defined role or training would behave more positively. If the initial cue is ambiguous, the most common positive response is to seek more information. People in groups will decide what to do depending on their perceptions of other people's reactions; without a dominant personality providing an early lead, people may take much longer to respond than if they were alone.

Once the existence of a fire has been recognised, people may undertake a wide range of possible behaviours [section 6-2]. These depend on role and training, and also people's perception of the developing fire situation (which may of course differ considerably from the truth). People may, for example, continue working, go to collect belongings, attempt to fight the fire, seek to warn others (role/training may dictate a search of the building; on the other hand, people may simply warn others they encounter), rejoin family groups, rescue/assist others, seek refuge or escape. As the perceived situation worsens, the range of viable options reduces [sections 6-6, 6-9].

The speed of movement of people [section 6-3], in crowded situations as well as when unimpeded, is well understood for most common situations. Empirical relationships have been derived for walking speed as a function of crowd density (people per square metre), flow rates in corridors or up/down stairs [section 6-5], and the rate at which queues of people pass through doorways [section 6-4] or other similar constrictions as a function of their width. The exit choices [section 6-8] made by different people may mean that not all doors are used by the optimum number of people to minimise the evacuation time. Some research has also been done to define the capabilities of disabled people [section 6-10], who will often move slower, and in some cases may find particular obstacles impassable. These relations enable the movement time portion of the total evacuation time to be estimated with a fair degree of accuracy [section 6-11]. However, this portion is often not the dominant component in the overall time it takes for evacuation.

As people move/remain within the building, they may be exposed to smoke and the toxic products of combustion [section 6-6]. The exposure is usually quantified in terms of a Fractional Effective Dose (FED) which depends on the concentration of particular toxins within the 'cocktail' of fire gases, and the duration of exposure. When the FED for a particular person reaches unity, he/she is overcome by the smoke. In the case of some combustion products (eg irritants), the effect is almost instantaneous rather than cumulative, so just the concentration is the key parameter. Tenability levels [section 6-7] for different rooms may be expressed in a number of ways, either the time for a person's FED to reach unity (if they remain in the room for the duration), or the value of the concentration of a given product that is sufficient to prevent escape effectively.

The total time for evacuation is termed the Required Safe Egress Time (RSET) [section 6-1]. This value is often compared with the Available Safe Egress Time (ASET) [section 2-2] to determine whether or not a particular scenario poses an unacceptable hazard.

User's notes

6 Human factors

Effect **of** human factors	Effect **on** human factors	
Fire growth ● First-aid firefighting may be effective in controlling or extinguishing the fire ● By opening or closing doors (or windows), people can affect the oxygen available to a fire	**Fire growth** ● Rapid increases in fire growth may be hazardous ● The non-linear growth of fire may give people a false impression of the time available to them ● People close to the fire may suffer from burn injuries	
Smoke spread ● By opening or closing doors (or windows), people can affect the movement of smoke around a building ● If doors to pressurised areas are opened for any length of time, the pressurisation will be lost	**Smoke spread** ● Toxic products of combustion, contained in smoke layers, are the primary cause of death/injury in fires. ● People may also be incapacitated by heat, either by immersion in smoke, or by radiation from hot layers above them ● Obscuration by smoke hinders or prevents egress ● Irritant smoke enhances the effects of obscuration	
Structural behaviour ● Leaving fire doors open may allow the spread of fire ● Fire resistance may be compromised by sloppy reinstatement after maintenance or modification of the building	**Structural behaviour** ● Structural collapse, while less likely to be a hazard for people (other than the Fire Service) inside the building, may nevertheless be a hazard to those outside	
Detection/Suppression ● Systems may be disabled during maintenance ● 'Responsible' staff may turn off alarms if they assume them to be false without investigating why the alarms have been set off	**Detection/Suppression** ● There will be no response until some form of detection/warning has occurred ● Early detection leaves more time available for escape ● Suppression will at least retard fire growth, leaving more time for escape ● Some systems (eg CO_2/Halon) can suffocate people in accidental discharge	
Fire Service intervention ● People may call the Fire Service to the fire	**Fire Service intervention** ● Arrival of the Fire Service may be the cue that persuades people that the fire is 'real' ● The Fire Service can rescue people unable to escape by themselves	

User's notes

Description

RSET is the time required after the fire has started, for the last person to reach a place of safety. It includes the time for:

- the fire to be detected,
- the alarm to be raised,
- people to recognise the alarm for what it is,
- people to respond to the alarm, and
- people to evacuate.

User's notes

Design issues

- RSET depends on the fire scenario chosen.
- When comparing RSET with Available Safe Egress Time (ASET), it is important to be clear which parts of the building are being considered.

Enforcement issues

- RSET and ASET are not single deterministic values, but will have probability distributions associated with them. Analysis must consider the safety factor, not just whether ASET > RSET.

References and resources

- *The SFPE handbook of fire protection engineering*, chapter 3-14
- *BS 7974: Application of fire safety engineering principles to the design of buildings.* (*Note:* RSET is referred to as 'escape time'.)
- *CIBSE guide E*, chapter 4
- Numerous evacuation models predict the movement of people: some examples are CRISP, Simulex, EXODUS, EVACNET, EXIT89, EvacSim, etc.

Links to other topics

Available Safe Egress Time [2-2], Egress modelling [6-11]

Further notes

The definition in terms of time since ignition is not universally accepted; some sources define RSET as the time since the first warning is given. What is important is that ASET and RSET are expressed on a common time basis.

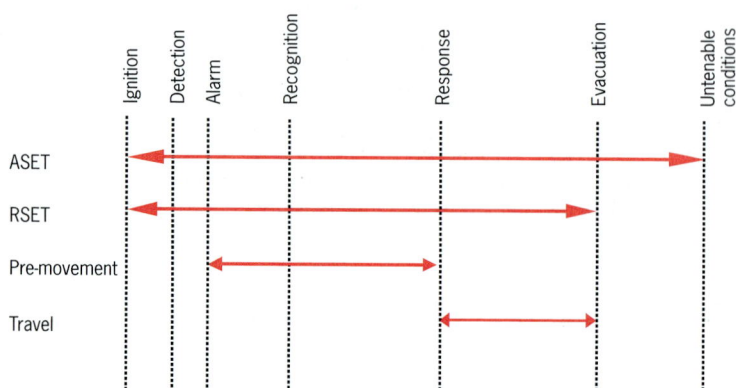

Pre-movement time

Description

The evacuation time is conventionally split into two components:
● travel time, and
● pre-movement time.
(*Note:* The Required Safe Egress Time (RSET) also includes a time for the fire to be detected and an alarm raised.)

The pre-movement time incorporates:
● the time to recognise that:
  ❏ an alarm has been given, and
  ❏ some action needs to be taken, and
● a 'response time' for all other activities performed prior to evacuation.

Design issues

● Pre-movement time depends on many different factors, eg type of warning given, people's activity at that time, their familiarity with the building and its systems, their behavioural roles and responsibilities, etc.
● The 'response' component may encompass many different activities, eg investigation, recognition of the threat, gathering family members, collecting valuables, searching the building, ordering others to leave, etc.
● A modelling short cut involves representing all pre-movement activities by a single time delay before a person moves; different people will move at different times. However, the probability distribution for this delay is hard to quantify without considering the activity explicitly.

Enforcement issues

● Pre-movement time can be a significant portion of the total evacuation time, so should not be neglected.

References and resources

● *BS 7974: Application of fire safety engineering principles to the design of buildings*
● *CIBSE guide E*, chapter 4
● Few evacuation models [section 6-11] predict the pre-movement phase (most use a distribution of delay times): some examples are CRISP, EXODUS (in development), EvacSim, etc.

Links to other topics

Available Safe Egress Time [2-2], Egress modelling [6-11]

Further notes

'Pre-movement time' includes activities involving movement.

User's notes

Input ◄──────► Output

Description

Not everybody moves at the same speed. Young adults walk faster than elderly people or children, and men walk faster than women. In uncrowded conditions, people can walk unimpeded; however, as the crowd density increases, movement slows down, and at some point effectively stops.

Design issues

● Walking speed is only one component of travel time, and travel time is only one component of the total evacuation time.

Enforcement issues

● As design

References and resources

● *The SFPE handbook of fire protection engineering*, chapter 3-14
● *CIBSE guide E*, chapter 4
● Evacuation models that treat each person individually allow specification of distributions of walking speeds

Links to other topics

Required Safe Egress Time [6-1], Available Safe Egress Time [2-2]

Further notes

As a rule of thumb, a default movement speed may be taken as 1.2 m.s^{-1}. Movement is unimpeded while the crowd density is less than 0.8 people.m^{-2}, and effectively zero above densities of 4 people.m^{-2}

User's notes

Input ◄────► Output

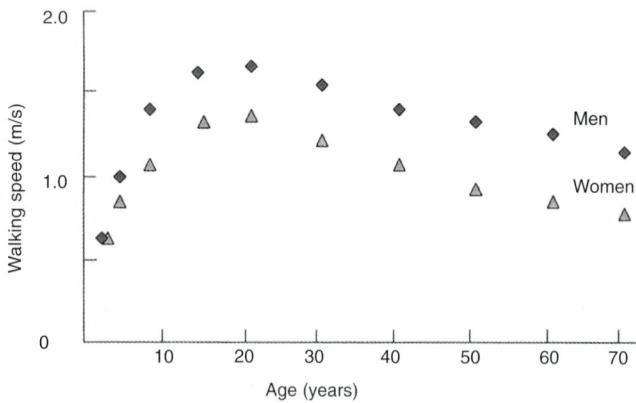

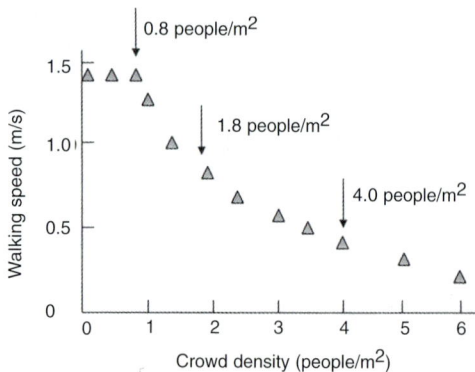

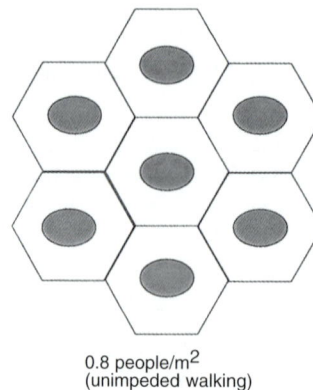

0.8 people/m^2
(unimpeded walking)

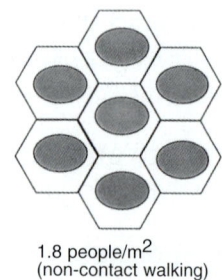

1.8 people/m^2
(non-contact walking)

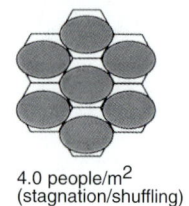

4.0 people/m^2
(stagnation/shuffling)

Movement through doors

Description
People move more slowly through doors than in the middle of a room, because they try to avoid contact with the edges of the door frame.

Input◀━━━▶Output

Design issues
● Once a queue has built up at a doorway, the time for the people in the queue to pass through the door may be more than the time required to move to join the queue from any point in the room. In this case, evacuation time depends primarily on the flow rate through the doors, and travel distances can be neglected.

Enforcement issues
● It should not be assumed that all doorways will receive just the right amount of traffic so that the last people to exit via each door do so simultaneously (optimising the evacuation time).
● In the absence of a more detailed analysis, the widest door should be assumed to be unusable as the consequence of a fire, to maximise the estimated time it will take for people to evacuate.

References and resources
● *The SFPE handbook of fire protection engineering*, chapter 3-14
● *CIBSE guide E*, chapter 4
● CRISP and Exodus account for reduced speed through doors, and presumably some other models do as well. In Simulex it arises naturally, from the movement algorithms used

Links to other topics
Egress modelling [6-11], Speed of movement [6-3], Exit choice [6-8]

Further notes
The speed of movement through doors is given by the equation:

$$S = 1.40 - 0.37d$$

where S is the speed (ms^{-1}) and d is the density (people.m^{-2}).

The maximum flow rate (people per second per metre of effective width) occurs at a density of just under 2 people.m^{-2} and has a value of 1.3 people.$s^{-1}.m^{-1}$.

The effective width is the actual clear width of the door, less a 0.15 m 'boundary layer' on either side.

Movement on stairs

Input ◄──► Output

Description

User's notes

People move more slowly on stairs than they do on the level. This is because they cannot take a natural stride, but are constrained by the geometry of the stair treads. Also, due to the risk of tripping (particularly when going down stairs), people only take a step when they have a secure foothold. As with movement on the level, the speed will be strongly affected by the crowd density (people per sq.m). Another issue is that it is considerably more tiring moving up or down, than moving on a level surface, hence people move more slowly to conserve energy, and stop to rest.

Design issues

● Optimum flow rates are achieved at crowd densities of about 2 people.m^{-2}. People's speed along the stair slope will be about 0.5 ms^{-1}; the horizontal component will be less and depends on the angle of the slope.
● Speeds up and down stairs are fairly similar.

Enforcement issues

● Stairs should not be considered as horizontal corridors, with their length adjusted to give the 'correct' time to move from one end to the other since this will only work for low crowd densities. It will not give the correct number of people who can occupy the stair, nor will it give the correct speed-density relationship.

References

● *The SFPE handbook of fire protection engineering*, chapter 3-14
● *CIBSE guide E*, chapter 4

Links to other topics

Egress modelling [6-11], Speed of movement [6-3]

Further notes

The evacuees' speed down stairs, as a function of crowd density, can be obtained from the following formula:

$$S = 1.08 - 0.29d$$

where S is the speed (ms^{-1}) and d the density (people.m^{-2}).

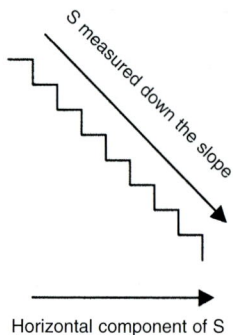

S measured down the slope

Horizontal component of S

6 6 Movement through smoke

Input ◄——► Output

Description

There are two issues here. The first is whether people will actually move (or continue to move) through smoke, or whether they will choose an alternative action/route leading away from the smoke. Generally, it is assumed that when the tenability limits are exceeded, further movement through smoke ceases. The second issue is the speed of movement; as visibility decreases, people will move more slowly. Irritant smoke has a greater effect in this respect than non-irritant smoke.

If people do move through smoke, they will be affected by exposure to the toxic products, heat, etc. This exposure is usually expressed in terms of a Fractional Effective Dose (FED); when this dose exceeds 1.0, a person is incapacitated.

Design issues

- A design approach that treats any encounter with smoke as a failure of the system may be too conservative.
- On the other hand, allowing people perfect knowledge to anticipate and avoid the movement of smoke will be too optimistic.
- People do not have the same FED uptake rates; the young/elderly are more susceptible than fit adults.

Enforcement issues

- As design

References

- *The SFPE handbook of fire protection engineering*, chapters 2-4, 2-6, 3-12

Links to other topics

Required Safe Egress Time [6-1]

Further notes

Such research as has been conducted on this topic has concentrated on when people are escaping. The probabilities of people moving through smoke (and/or associated tenability limits) for different actions such as investigation, first-aid firefighting, warning or rescuing others, etc, have not been determined.

Most egress models neglect the design issues above, however some are addressed by EXIT 89, CRISP and EXODUS.

User's notes

Description

Tenability limits are the properties of the smoky environment, when conditions are sufficiently bad to prevent escape from/through a compartment/egress route. Some limits just depend on the concentration (in the loosest sense) of a combustion product (eg irritants) or smoke optical density since these 'instantly' prevent escape when the critical value is reached. Other limits depend on concentration and likely time of exposure, eg carbon monoxide, heat. These are often related to a FED; when FED exceeds 1.0 a person is incapacitated.

User's notes

Design issues

● There are no standard values for tenability limits for particular combustion products.
● In some cases (eg optical density affecting loss of visibility), the limits may depend on the occupants' familiarity with the building, ie represent willingness to move through smoke, rather than inability to see.
● Better visibility is required for larger rooms.

Enforcement issues

● If the smoke layer is stratified above a clear layer, the layer interface height should also be one of the tenability criteria. Some safety margin should be allowed, ie the limit should not be set at 'head height' or below.
● Layer height is not sufficient to cause loss of tenability; one of the other limits relating to smoke properties must also be exceeded.

References and resources

● *The SFPE handbook of fire protection engineering*, chapter 2-6
● Numerous simulation models predict the movement of smoke and the dilution of combustion products: some examples are JASMINE, CFX, FDS, CFAST, CRISP, etc.

Links to other topics

Available Safe Egress Time [2-2], Required Safe Egress Time [6-1]

Further notes

Although there are no standard values, the following (from the SFPE Handbook) may be taken as rules of thumb for the common parameters affecting building occupants:

Carbon monoxide	6000 ppm
Carbon dioxide	7%
Oxygen deficiency	13%
Hydrogen cyanide	150 ppm
Irritants	1 ppm (acreolin); 75 ppm HCl
Smoke optical density	0.1 m^{-1}
Temperature	80 °C
Radiant heat flux	2.5 kW.m^{-2}
Clear layer height	1 m above head (for small rooms/corridors

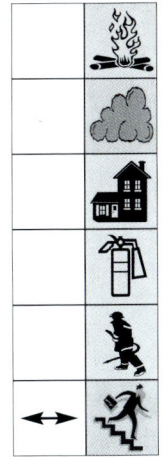

Exit choice

Description
The factors that influence exit choice are complex, and quantitative data is scarce. In qualitative terms, people tend to stick to familiar routes. A distinction can therefore be made between buildings where the occupants are familiar with the geometry (eg office workers), and buildings where they are not (eg public assembly). Even in the former types of buildings, it is good design to ensure the fire exits are part of the normal circulation routes. A clearly-signed 'Fire Exit' may be ignored in preference to the familiar route. However, building visitors given directions by building staff are more likely to choose unfamiliar exits.

Design issues
● The location of the fire may make (at least) one exit inaccessible. Alternative fire locations may need to be tried to identify the 'worst case'.

Enforcement issues
● People will not necessarily choose their nearest exit.
● People will not choose exits on the basis of their width, ie the number of people using each exit will not be optimised to minimise the total evacuation time.
● As it is hard to quantify the probability of choosing any given exit, sensitivity studies should be performed.

References and resources
● **Canter D A.** *Fires and human behaviour: psychological aspects of the experience of fires.* London, David Fulton Publishers, 1990
● A number of evacuation simulation models attempt to address the issue: some examples are CRISP, EGRESS, E-SCAPE, EXIT89, EXODUS, SIMULEX, STEPS

Links to other topics
Required Safe Egress Time [6-1]

Further notes
In most evacuation simulation models, the length of the shortest route to the 'outside' (or place of refuge) is calculated for every possible location a person may occupy. The direction each person moves in is then chosen to minimise continually the distance they still have to go.

Exit choice is built into this modelling mechanism by adding additional distance for some possible routes from each location, but not for other routes. The effect is to prevent unfamiliar exits being chosen until a person is much closer to an unfamiliar exit than a familiar one.

Since quantitative data is scarce, it is likely that the additional distances are chosen to determine the modeller's 'best guess' of what the probabilities of exit choice are.

User's notes

Input ←——→ Output

Input ◄─────► Output

6
9   Panic: a myth

Description
People rarely panic in fires. They choose from a range of
behaviour options depending on their assessment of the prevailing
conditions. However, it must be recognised that in some cases,
their assessment may be incorrect, based on inadequate or
ambiguous information. With the benefit of hindsight, it is easy to
dismiss the person's actions as 'irrational' or 'panic'.

User's notes

Design issues
● 'Panic' cannot be invoked as a reason to enable people to move
  more quickly.

Enforcement issues
● As design

References
● **Canter D A.** *Fires and human behaviour: psychological aspects
  of the experience of fires.* London, David Fulton Publishers, 1990

Links to other topics
Movement through smoke [6-6], Exit choice [6-8]

Effect of disabilities

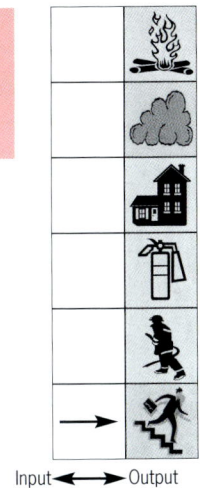

Input ◄——► Output

Description

There are many types of disability, physical, sensory or mental, and different degrees of severity. There are also many adverse impacts that disabilities may impose on a person's ability to be aware of, react to and escape from a developing fire situation.

Design issues

- Sensory disabilities may negate any benefits of early detection and alarm, or prevent recognition of deteriorating conditions.
- Mental disabilities may prevent appropriate responses to cues and conditions.
- Physical disabilities may affect speed of movement, and prohibit certain routes being used unaided (eg wheelchairs on stairs, insufficient strength/dexterity to open a self-closing door, need for frequent stops to rest).
- Disabled people may not be accompanied by able-bodied people who can give assistance.

Enforcement issues

- Special fire safety precautions may be required to handle the needs of disabled people.

References and resources

- *The SFPE handbook of fire protection engineering*, chapter 3-12
- **Shields T J.** *Fire and disabled people in buildings*. BRE report BR 231. 1993
- **Shields T J, Dunlop K E & Silcock G W.** *Escape of disabled people from fire: a measurement and classification of capability for assessing escape risk*. BRE Report BR 301. 1996
- A number of models do allow a 'disabled sub-population' to be generated with slower movement speeds than normal. At present, no egress model explicitly takes account of other effects of disabilities.

Links to other topics

Required Safe Egress Time [6-1], Speed of movement [6-3]

Egress modelling

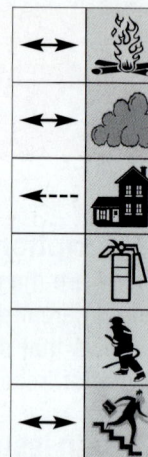

Description

Numerical models of the evacuation process vary widely in their degrees of sophistication. The simplest treat the population as a homogeneous fluid, or mindless particles, and concentrate on the flow capacity of the building. At the other extreme, there are detailed simulations where each person is treated individually, with explicit behavioural rules.

Input ◀━▶ Output

Design issues

- Building geometry can either be described by a coarse network (each node ~ 1 room) or a fine network (each node ~ 0.5 m 'pixel'). Fine networks are necessary to calculate complex flow rates from first principles; coarse networks would require empirical equations.
- The more 'homogeneous' the population behaviour, the less distinct the difference between coarse and fine networks (ie all flows heading to exits, no contraflows and limited merging).
- Simpler models are easier to use and faster to run. However, realistic behaviours are required for fully realistic results.
- Some models allow interaction with the fire environment [calculated by a fire model, either a-priori (most cases) or at the same time as in the evacuation model, CRISP].

Enforcement issues

- Human behaviour is the most complex and difficult aspect to simulate, yet is crucial to accurate results.
- Not all aspects of behaviour are fully understood or quantified yet, so sensitivity analysis is important.
- Validation of the behavioural aspects of existing models is currently rather poor.

References and resources

- *The SFPE handbook of fire protection engineering*, chapter 3-14
- *CIBSE guide E*, chapter 4
- **Gwynne S, Galea E R, Owen M, Lawrence PJ & Filippidis L.** A review of the methodologies used in the computer simulation of evacuation from the built environment. *Proceedings of 1st Human Behaviour in Fire Conference*, 1998. p 681
- **Fraser-Mitchell, J N.** Modelling human behaviour within the fire risk assessment tool CRISP. *Fire & Materials* 1999: **23:** 349–355
- CRISP, EXODUS, SIMULEX, EXIT89, EVACNET

Links to other topics

Speed of movement [6-3], Required Safe Egress Time [6-1]

7 Risk assessment

Fire safety engineering has been defined as:

'The application of scientific and engineering principles based on an understanding of the phenomena and effects of fire and of the behaviour of people to fire, to protect people, property and the environment from the destructive effects of fire.'

The principal objective of fire engineering is, when an accidental fire occurs, to provide an environment that has an acceptable level of safety. Often this will involve calculation or modelling [sections 1-9, 3-7, 6-11] of scenarios affecting all or part of the fire 'system'. Implicitly or explicitly, a form of risk assessment is involved. It is important to match the right method to the decision to be made.

Implicit risk assessment examples include the comparison of calculation results with threshold criteria, for example 'smoke layer well above people's heads' or 'area of fire spread restricted to less than X sq.m'; often these are linked with 'worst case' scenarios [section 7-8]. The idea is that 'worst' and lesser scenarios have minimal consequence, other more severe scenarios being assumed to have minimal probability. How do you know which scenario will be 'worst case'? Assumptions which might be 'conservative' for one aspect of the fire 'system' might not be conservative at all for other aspects. A sensitivity analysis [section 7-1] should be performed to estimate the consequences of uncertainties in the scenario, variable values, etc.

Explicit risk assessment uses the formula 'risk = probability x consequence', summed up for all hazards. Every fire safety decision should require a full risk analysis, until or unless it can be shown that a less comprehensive approach is adequate. The preferred approach to uncertainty is to quantify it, rather than rely on conservative assumptions.

No building can be completely safe, yet there is an unresolved question of what absolute level of risk should be deemed acceptable. A further problem is that acceptable risk varies with circumstances and public perception. There are two ways in which these problems may be addressed.
- Assume that the risks associated with buildings constructed following prescriptive guidelines are 'reasonable'.
- Use quantitative methods in comparative mode, ie assume that systematic errors and biases 'cancel out' when two similar buildings are compared.

Absolute risks are not calculated since the state of the art is insufficient to do this without a large degree of uncertainty.

Complexity	Fire risk method	Subjectivity
	Elementary methods [section 7-2]	
	Points schemes [section 7-3]	
	Statistical/Probabilistic methods [sections 7-4, 7-5]	
	Event trees [section 7-6]	
	Monte-Carlo simulation [section 7-7]	

User's notes

7
1

Sensitivity analysis

Input ◄──────► Output

Description

The outcome of a fire engineering calculation will frequently depend on many variables, all of which have some degree of uncertainty in their values. A sensitivity analysis involves comparing the outcomes of many calculations, where each pair of calculations differs only in the value of one of the variables.

Design issues

● The range of values that should be tried for a given variable should depend on the level of uncertainty.
● If the outcome is not sensitive to the value of a given variable, it does not matter if that variable's value has large uncertainty. On the other hand, if the outcome is sensitive, then measures must be taken to reduce the uncertainty (even if it is already 'small').
● The outcome may be a highly non-linear function of the variable values, so just looking at two or three possible values for a variable may miss regions where the sensitivity is high.

Enforcement issues

● Sensitivity analysis is essential.

References

● *The SFPE handbook of fire protection engineering*, chapter 5-9
● Many textbooks exist on statistics and experimental design, which will cover sensitivity analysis

Links to other topics

Safety factors [7-5]

Further notes

It is not necessary to consider every possible combination of different values. If the outcome O depends on a number of independent variables x, y, …, then by Taylor's series expansion:

$$O(x + \Delta x, y + \Delta y,...) = O(x, y,...) + \left(\frac{dO}{dx}\right)\Delta x + \left(\frac{dO}{dy}\right)\Delta y +...$$

The differentials can be calculated thus:

$$\frac{dO}{dx} = \frac{O(x + \Delta x,...) - O(x - \Delta x,...)}{2\Delta x} \quad \text{as } \Delta x \to 0$$

Similarly for the others.

The change (uncertainty) in the outcome will be large if either the differential (dO/dx) is large, or the uncertainty in the variable value (Δx) is large. As there is nothing that can be done to alter the value of the differential (the 'sensitivity'), the only way to reduce the uncertainty in O(x, …) is to reduce the uncertainty in x.

(Also, watch out for singularities where dO/dx is infinity!)

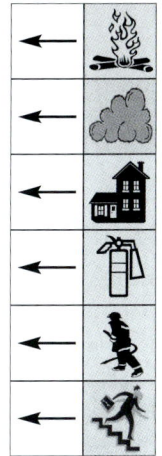

Description

At the simplest level, in elementary methods, hazards may be defined as low, medium or high severity, with probabilities of occurrence similarly assessed as low, medium or high. If most or all contributions to the risk are 'low severity x low probability' then a more rigorous approach may be unnecessary.

User's notes

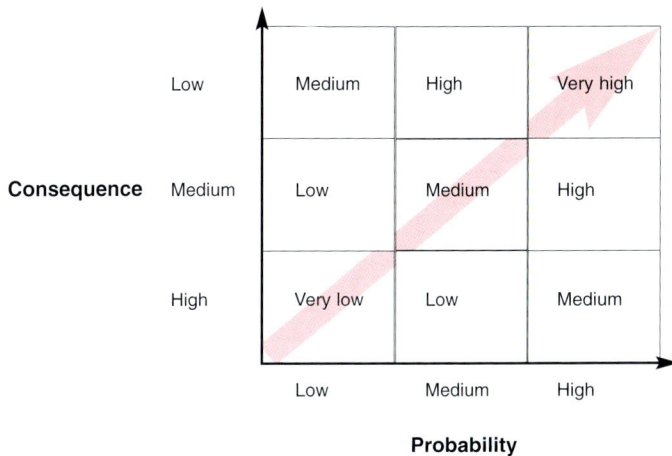

Probability

Design issues

● Elementary methods are best used to give a rapid overview of the system prior to a more in-depth investigation of key areas, particularly where potential consequences are severe.

Enforcement issues

● Such methods can enable routine risk management on a day-to-day basis (eg to check that good practice is being maintained).
● Presence of a risk assessment, however crude, can be taken as evidence of a 'safety culture' and may be sufficient for legal purposes.

References

● **Fire Protection Association.** *An Introduction to the Fire Precautions (Workplace) Regulations* (2001)
● **Kazer B M.** Risk assessment — a practical guide. *Safety & Practitioner* 1993 (May): Supplement
● **Everley M.** Why risk it? *Health & Safety at Work* 1994 (Nov): 34–36

Points schemes

Description

Points schemes calculate a simple numerical score, based on parameters such as size of the building, fire load, presence of sprinklers, alarms, etc. Risk is related to the score in some way. There are various sub-classes of points schemes but really only two significantly different types:

- linear or 'pure' points schemes where various factors are awarded points, multiplied by weighting factors, and the final total compared with a pass/fail threshold.
- factorial schemes, which express the probable level of safety as a ratio of the overall hazards to the protective and prevention measures. These factors may be represented by fairly complex non-linear equations.

Design issues

- Points schemes may be reasonably effective at ranking different buildings according to their relative risk, but do not give absolute or even relative risk values.
- Most points schemes, especially the simpler forms, have an inability to handle the interactions between different fire safety systems.

Enforcement issues

- Methods do not seem to have been calibrated beyond 'they appear to work'. Scoring methods may not be readily translatable to other risk assessment approaches
- Subjectivity enters into the process when the user of the scheme must assign numerical values to various parameters. Some schemes help the user by restricting his input to 'yes/no' answers to the list of questions. However, other parameters eg 'quality of staff' may be much more difficult to quantify.

References and resources

- **National Fire Protection Association.** *Guide on alternative approaches to fire safety.* NFPA 101A. Quincy, MA, NFPA.1995
- *Buildings and fire,* chapter 13
- **Stollard P.** The development of a points scheme to assess fire safety in hospitals. *Fire Safety Journal* 1984: **7:** 145–153
- **Kaiser J.** Experiences of the Gretener method. *Fire Safety Journal* 1979–1980: **2:** 213–222
- **De Smet E.** *FRAME manual,* version 2
  see also http://user.online.be/~otr034926/webengels.htm
- NFPA 101A contains a paper-based points scheme (FSES).
- FRAME is a cheap computerised method, derived from the Gretener scheme

Links to other topics

Statistical methods [7-4]

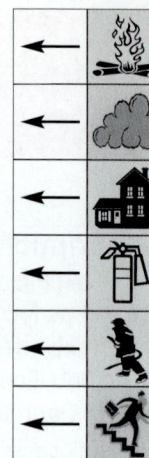

User's notes

Input ◄——► Output

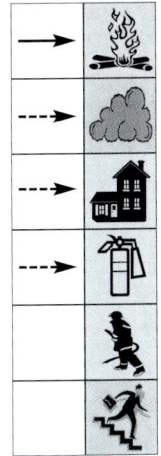

Statistical methods

Description

Fire statistics are the nearest analogy we have to direct experiments on the whole fire 'system'. However, unlike true experiments, the large range of possible variables cannot be investigated in a controlled manner. It is therefore sometimes difficult to determine which are the true dependent variables.

User's notes

Input ◄────► Output

Design issues

- Statistical approaches to risk assessment can only be used for existing types of buildings. Alternatively, statistics can be used to estimate probabilities at the component level.
- Where suitable statistics are available, their simplest application is in the direct estimate of probabilities. This may include the derivation of probability distributions as well as point estimates.
- Regression techniques may be used to derive empirical relationships between variables (eg risk of death or injury, depending on the area damaged by fire).
- The forms of various distributions may not be known, but instead may be assumed to be of a certain type for mathematical convenience. Skewed distributions are frequently assumed to be Log-Normal, or Weibull. When a distribution depends on many random variables, the distribution may be assumed to be normal.

Enforcement issues

- Fire statistics may form a biased sample (from the 'population' of all fires occurring) owing to the way the statistics are collected (ie Fire Service reports on fires attended). For instance, in the UK it is estimated that the Fire Service only attends about 15% of fires.

References and resources

- *The SFPE handbook of fire protection engineering*, chapter 1-12
- Numerous textbooks on statistics and probability are available
- UK Home Office Fire Statistics
- NFPA 'one stop data shop'

Links to other topics

Safety factors [7-5], Points schemes [7-3], Logic trees [7-6], Monte Carlo simulation [7-7]

Safety factors

Description
The assessment of a design requires the comparison of the value of a design variable, D, with a pass/fail threshold value, t. The *safety margin*, M, will be the difference (D − t); the design fails if M < 0.

Due to uncertainties in the calculations and their values, the design variable D will be defined by a probability distribution with a mean value μ_d and a standard deviation σ. The safety margin will therefore also be a random variable, this time with mean value $\mu_m = (\mu_d - t)$, and the same standard deviation σ.

The safety index β is defined as $\beta = \mu_m / \sigma$.

The value of β can be used to calculate the probability that the design will fail the acceptance criterion, ie pr(M < 0).

Design issues
● The uncertainty in the value of the design variable (expressed as the standard deviation, σ) is much harder to quantify than the mean value μ.

Enforcement issues
● *Note:* the safety margin is a multiple of the standard deviation of D, not the mean value of D.

References and resources
● BS 7974 PD 7: *Probabalistic risk assessment*
● **Magnusson S E, Frantzich H, Harada K.** Fire Safety Design based on calculations: uncertainty analysis and safety verification. *Fire Safety Journal* 1996: **27:** 305–334
● Tables of the normal distribution can be found in almost any textbook on probability and statistics

Links to other topics
Statistical methods [7-4]

Further notes
It is frequently assumed that D (and hence M) is a normally distributed random variable. This is usually justified by the Central Limit Theorem if D is in turn a function of several random variables.

User's notes

Input ◄───► Output

Probabilities of failure for a normal distribution for selected values of β	
β	pr(fail)
0	0.5
1.0	0.15
1.645	0.05
2.237	0.01
3.09	0.001
3.7	0.0001

Logic trees

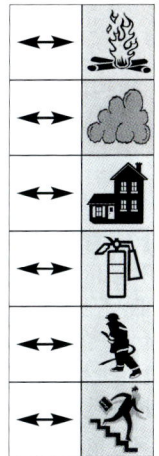

Input ←→ Output

Description

Event trees are inductive logic methods for identifying the various possible outcomes of a given initiating event. Fault trees are closely related, looking at the various sequences of events which could lead to a particular outcome.

The evolution in time of the total fire 'system' is described in terms of transitions between discrete states. At each point where an event may occur, there is a branch between possible future outcomes. Analysis of the entire probability tree (formed by all these branches) allows the probability to be calculated for all possible outcomes.

Design issues

- The consequences of the scenarios require either modelling or testing to evaluate them; then we can evaluate the risk as the sum of {probability × consequence} for all scenarios.
- The probability of an event, as mentioned previously, may be dependent on many factors, including time. This may greatly increase the tree's complexity.
- Handling of rare events is no different to common events.
- Fault trees are more geared to component-level analysis, whereas event trees have a broader outlook.

Enforcement issues

- Despite the claim that event trees are comprehensive and examine all possible outcomes, in practice this may only be achieved if the number of events (and hence branches of the tree) is strictly limited.
- Due to uncertainties in all the input parameters, the accuracy of absolute risk estimates is only about an order of magnitude.

References and resources

- *The SFPE handbook of fire protection engineering*, chapter 5-12
- **Fitzgerald R W.** An engineering method for building firesafety analysis. *Fire Safety Journal* 1985: **9:** 233–243
- **Beck V R.** Performance-based fire engineering design and its application in Australia. *Proceedings 5th IAFSS Symposium on Fire Safety Science*, Melbourne, Australia, 1997. pp 23–40
- **Yung D, Hadjisophocleous G V & Proulx G.** A description of the probabilistic and deterministic modelling used in FiRECAM. *International Journal of Engeering Performance-Based Codes* 1999: **1**(1): 18–26
- Fitzgerald's METHOD, FiRECAM, FIERAsystem, Cesare-RISK

Links to other topics

Statistical methods [7-4], Monte Carlo simulation [7-7]

Cont'd....

Worst case scenario

Description

In a deterministic study (as opposed to a full probabilistic risk assessment), initial assumptions about fire size and growth rate, numbers of building occupants, etc., are usually sufficiently conservative so as to define a 'reasonable worst case' scenario. The consequences of this scenario can be calculated and compared with the assessment threshold criteria (eg 'smoke layer well above people's heads' or 'area of fire spread restricted to less than x m²'). A system designed to handle a 'worst case' scenario is assumed to perform acceptably in less severe scenarios; more severe scenarios are assumed to have minimal probability.

User's notes

Design issues

- The consequences of the scenarios are not known before they are calculated so it may be necessary to define several 'worst case' candidates, and evaluate each one in turn.
- Using extreme initial assumptions will be over-conservative (ie very low probability of this scenario ever happening). On the other hand, using 'typical' or average values will not be conservative enough.
- The scenarios must be self-consistent; for example, a limiting fire size in a sprinklered building would no longer be appropriate if the sprinklers were to fail.
- A sensitivity analysis [section 7-1] will estimate the consequences of uncertainties in the scenario, variable values, etc.

Enforcement issues

- Assumptions which might be 'conservative' for one aspect of the fire 'system' might not be conservative at all for other aspects.
- A large, rapidly growing fire is not necessarily the most hazardous: an undetected smouldering fire in a closed room could kill a sleeping occupant.
- The 'worst case' may not be defined by a big fire with all fire safety systems working as designed. If a component of the system (eg detection, suppression, smoke control) fails, the consequences might be severe enough to outweigh the smaller probability of this scenario occurring.

References

- BS 7974: *Application of fire safety engineering principles to the design of buildings*
- *CIBSE guide E*, chapter 2

Links to other topics

Qualitative design review (QDR) [G-6], Assessment of designs [G-7], Risk assessment [section 7]

Further notes

A deterministic study does not attempt to quantify the probabilities of scenarios so the selection of scenarios likely to contribute the greatest {probability × consequence} to overall risk is subjective.

8 Greek alphabet

Scientists and engineers are fond of using Greek letters in their equations. For those who are not familiar with these characters the Greek alphabet is listed below, including a guide to pronunciation.

Letters of the Greek alphabet

Lowercase	Uppercase	Name	Pronunciation
α	A	alpha	al-fuh
β	B	beta	bee-tuh
γ	Γ	gamma	gam-uh
δ	Δ	delta	del-tuh
ε	E	epsilon	ep-sil-on
ζ	Z	zeta	zee-tuh
η	H	eta	ee-tuh
θ	Θ	theta	thee-tuh
ι	I	iota	eye-oh-ta
κ	K	kappa	kap-uh
λ	Λ	lambda	lam-duh
μ	M	mu	mew
ν	N	nu	new
ξ	Ξ	xi	ks-eye
o	O	omicron	om-i-kron
π	Π	pi	pie
ρ	P	rho	row
σ	Σ	sigma	sig-muh
τ	T	tau	tau
υ	Y	upsilon	oop-si-lon
φ	Φ	phi	fie
χ	X	chi	k-eye
ψ	Ψ	psi	sigh
ω	Ω	omega	oh-may-guh

9 References

Common references

BS 7974: *Application of fire safety engineering principles to the design of buildings.* Part 0: Guide to design framework and fire safety engineering procedures

Drysdale D. *An introduction to fire dynamics.* Second edition. Chichester, John Wiley. 2000.
ISBN 0 471 97290 8 (ppc) 0 471 97291 6 (pbk)

Morgan H P et al. *Design methodologies for smoke and heat exhaust ventilation.* BR 368. Garston, BRE Bookshop. 1999. ISBN 1 86081 289 9

Society of Fire Protection Engineers. *The SFPE handbook of fire protection engineering.* Third edition. Boston, Massachusetts, SFPE. 2002.
ISBN 0 87765 451 4

Buchanan A H. *Structural design for fire safety.* Chichester, John Wiley. 2001. ISBN 0 471 88993 8, 0 471 89060 X (pbk)

Shields T J & Silcock G W H. *Buildings and fire.* Harlow, Longman Science and Technical. 1987. ISBN 0 470 207750 7

Quintiere J. *Principles of fire behaviour.* Florence, Kentucky, Delmar. 1997. ISBN 0-8273-7732-0

Purkiss J A. *Fire safety engineering design of structures.* London, Butterworth-Heineman. 1996. ISBN 0 750 60609 6

Other references

Full details are given for other references that only have specialised relevance, where they are cited in the text.

Computer models

Full references to computer models have not been given since new models are continually being developed. The web site www.firemodelsurvey.com gives further details of models mentioned in the text and new models.

10 Index

Index

Index